please JOIN

THERE'S NO PLACE LIKE TECH

Copyright © 2024 by Debra Christmas and Kelley Irwin

All Rights Reserved.

The authors have made every effort to ensure the accuracy of the information within this book was correct at time of publication. The authors do not assume and hereby disclaim any liability to any party for any loss, damage, or disruption caused by errors or omissions, whether such errors or omissions result from accident, negligence, or any other cause.

please JOIN: THERE'S NO PLACE LIKE TECH

ISBN 978-1-7773018-6-6 (PB)

ISBN 978-1-7773018-5-9 (Digital)

Cover & Book design by Doris Chung

Published in Canada

FIRST EDITION

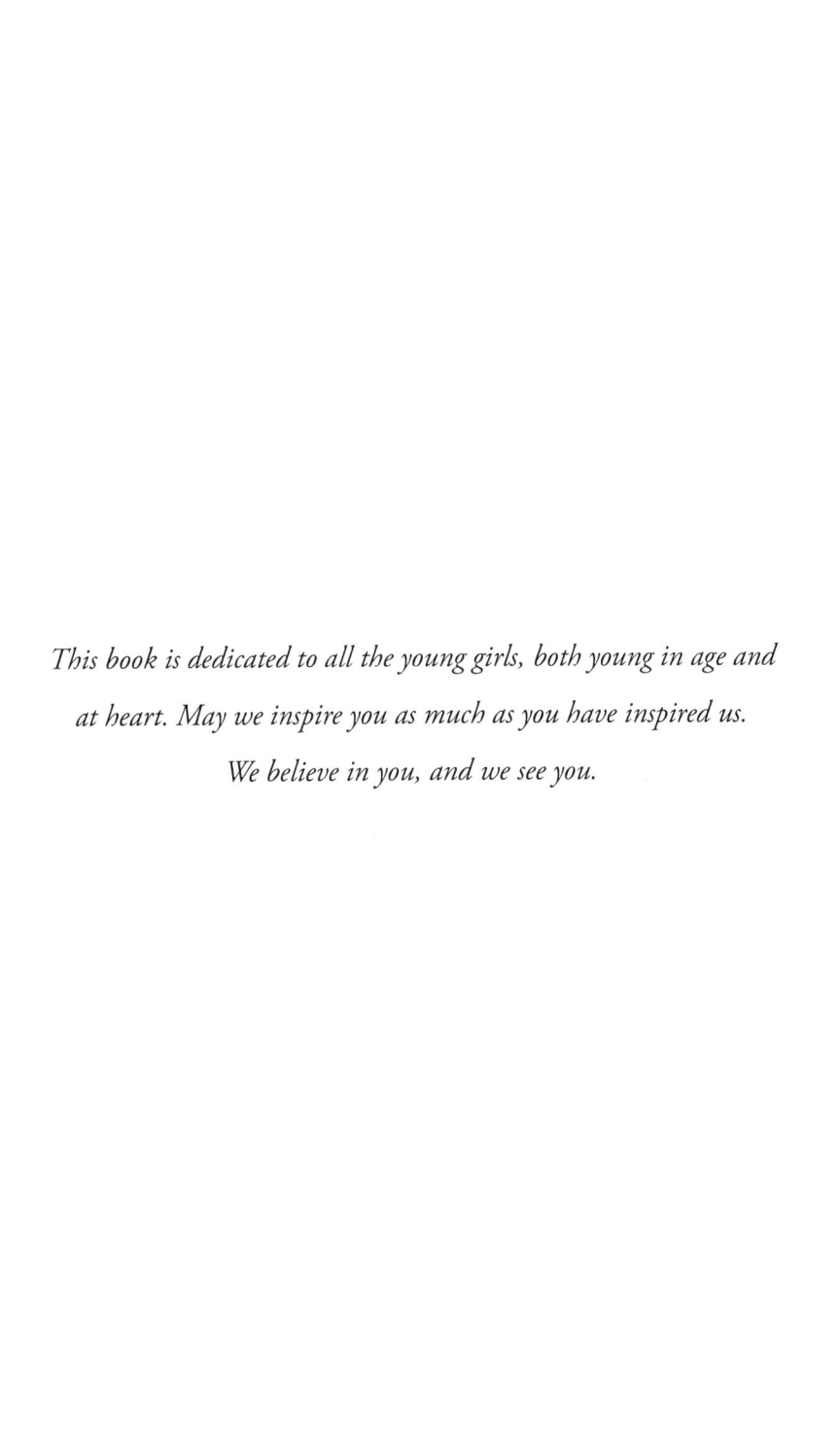

This book is dedicated to all the young girls, both young in age and at heart. May we inspire you as much as you have inspired us. We believe in you, and we see you.

Please JOIN

THERE'S NO PLACE LIKE TECH

Debra Christmas *Kelley Irwin*

DEBRA CHRISTMAS and KELLEY IRWIN

Table of Contents

Introduction	1
Please Join	5
Strength	13
Courage	35
Independence	59
Leadership	83
Assertiveness	103
Competitiveness	125
Perseverance	149
Confidence	171
Final Thoughts	195
Resources	199
Acknowledgements	211
About the Authors	217
Please Join Us	223
Endnotes	225

Introduction

Women in tech have the right to succeed and the responsibility to help others.

We published our first book in 2020 focused on women in tech—*Please Stay: How Women in Tech Survive and Thrive*. This book was a commitment from us to support women in technology careers and encourage them to remain in the field.

Once the book launched, we were asked what we could do for the girls. How do we encourage girls to enter the tech field, how do we support the parents and teachers in their lives, and could we offer practical advice and inspiring stories, and not just the statistics on the small number of girls entering the field?

From those conversations and requests, we decided to speak to girls to get their thoughts. We interviewed girls between the ages of 6 to 17, felt their intense energy and gained perspective on their futures. This enabled us to discuss the path forward with their parents.

These girls inspired us to take on the challenge and write about their incredible accomplishments and barriers they encounter. Our goal was to profile the talents of these girls, inspire other girls with their stories, and offer tips and techniques for girls, their parents and teachers to use along the way.

We also heard clearly that while not all girls want to enter a tech career, they understand they will be utilizing technology in every aspect of their lives. They want to be knowledgeable, they want to be courageous, and they want to be confident as they take on new challenges.

We believe in them.

We support them.

We were them.

Please Join

This is the time.

Women working in the Information and Technology (IT) sector make up less than 28% of the global IT workforce[1] in 2023.

Girls who are nurturing and caring are often encouraged to become nurses. Let's broaden this thought for these girls. Nurturing and caring are also strengths valued in technology careers. Consider, in addition to tech innovation, the design discussion which is centred on human needs.

Girls who are interested in educating others are encouraged to become teachers. Let's extend the options. This interest is a tremendous strength in technology careers. Technologists think about the people who will benefit from their product. They consider the user experience and work to minimize the training required to embrace its use.

Girls who are friendly and pay attention to detail can become administrative assistants and office managers. These roles are critical to keeping the workplace productive. Let's expand the choices. Being friendly, collaborative and detail-oriented are strengths that benefit every technology team as they consider options to solve problems and take all ideas into account to incrementally improve software products.

Girls who listen well may be encouraged to become social workers. Let's take this a bit deeper. This trait is also important in tech careers as team members benefit from hearing different opinions and ideas before committing to the path ahead.

Girls who are analytical and persuasive through logical reasoning are encouraged to become lawyers. Let's explore the possibilities. Analytical and persuasive strengths are dynamite in a technology career because you can influence the technical outcomes.

When was the last time you observed a young girl demonstrating a strength and heard someone say, "You should be in tech!"?

This is the time.

When did you personally see a young girl figuring out a problem and say to her, "You should be in tech!"?

This is the time.

Let's open their eyes, broaden their thoughts and encourage their paths.

Girls may not know how their interests and skills could apply to a tech career. We can tell them.

Girls may not understand the power of jobs for people who write code. We can tell them.

Girls may not know there are jobs in tech beyond people who develop code. We can tell them.

Girls may not know the pleasure of innovating, building and launching a tech product that enhances the employee or customer experience. We can tell them.

Girls may not know there was a time we didn't have the internet, digital cameras, laptops or mobile phones. We didn't have the ability to search for information online, use real-time navigation (GPS), or tap and pay for purchases. Each of these innovations became real and useful and ingrained in our lives after a technology team created them. We can tell them.

Girls opt out of Science, Technology, Engineering and Math courses for many reasons. When they do, it impacts potential career opportunities easily available for them to explore. Let's challenge them when they want to opt out and find out what's really happening. Let's encourage them. Let's support them on their path to success.

This is the time.

Girls are watching us. And listening to what we tell them.

We could fill an entire book with the issues including "society teaching girls to focus on perfecting rather than building, abiding by rules rather than breaking them".[2]

But we are better than that.

And girls deserve more than that. We need to openly talk about careers in tech, what they really feel like, what skills lead to excelling in those careers, and how girls can prepare for those opportunities in an industry where they can make a difference.

We need to talk about how they can explore, how they can be supported, and how they can become role models for other girls who may also be hesitant when they think of a career in technology.

Girls know that technology changes the world. We need to help them see their ability to be a part of technology and innovation. They can change civilization and find success and joy as part of this field. It's not just about coding, or even Computer Science, as they are both only a part of what information and technology is today. The skills and aptitude that we need far surpass a degree in Math or Science. Technology teams require creative people, innovative thinkers, excellent problem solvers, and people with social and communicative skills. Excellent technologists understand both the human needs and the technical possibilities to create wonderful, usable, feature-rich solutions for real life and real people. They are thoughtful, caring, talented and energized.

We need girls from all walks of life, with a wide range of knowledge and competencies to choose technology as a career. More importantly, these girls need to know we are here for them, we are counting on them, and we believe in them. We need them to **please join.**

We need their energy, their ideas, their power, and their unique gifts to bring IT to life in ways we have never seen before. **Please join.**

We profile stories, people, methods and resources to help you encourage the girls in your life. You can help them find their strength, their courage and their independence. You can encourage them to demonstrate their assertiveness and competitiveness in ways that feel natural to them.

Please Join is about leveraging the innate strengths and interests in the area of their personal choice. Girls have much to offer the tech industry and can experience a successful and promising future as tech continues to evolve. This will take conscious thought from parents and teachers, and actions from the girls and the other adults in their lives. It will require us to be deliberate around the world in all industries and all communities, to talk about tech, to encourage girls to consider tech, support them as they explore the profession and let them know they can excel.

The good news—girls with tech skills are all around us. There are talented girls who belong in IT, who will be allies as they connect with each other, and who will enjoy being part of a meaningful community where they can make a difference. We just need to encourage them along the way.

The future is bright for girls to enter tech. We need them and we are here for them so let's encourage them to **Please join!**

Strength

Strength–intellectual strength is the capacity of an individual to learn new things in the environment and apply them.

Girls and boys have the capacity to build intellectual strength. It's important to recognize these skills, learn how to increase them and practise them. The diverse types of intellectual strengths include analysis, problem-solving, verbal comprehension and reasoning skills.

Learning these skills can be fun and rewarding. Fun because they can be learned through activities in your daily life. Activities you enjoy, that can be done alone or in a group of friends, and can be accomplished in short bursts or in longer, more complex scenarios. It's important that you don't give up too quickly.

Analysis is a skill that allows you to leverage critical thinking, make tough decisions and solve problems. You take in information, process it, add to it experiences that you already have, make conclusions and identify options.

Problem-solving is a skill that allows you to analyse and evaluate information. You can look at scenarios that are completely new to you, think about the information that is available, identify potential options and test them out.

Verbal comprehension is a skill that allows you to take in spoken or written words, consider the meaning of each of the words individually as well as the way they relate to and interact with each other. You can assess your understanding by repeating back the words and their meaning with another person to see if there is a different interpretation to consider.

Reasoning is a skill that includes how you think when you take in information, analyse, evaluate and synthesize all of it with your creative and abstract thinking. You consider the context, the possibilities and the intent of the messages, and not just the written or spoken words.

We have a responsibility as we raise girls to provide opportunities for them to learn and practise new skills in real-life situations.

DID YOU KNOW?

The WIT Network is a global and local not-for-profit organization providing inspiration, education, mentoring, networking and practical advice to empower women (and men) to build and grow their careers in technology and pursue their ambitions. There are eleven thousand members from 76 countries, and 40 local communities located in Asia-Pacific (APAC), Canada, Europe, the Middle East and Africa (EMEA), and the United States (USA). There are both individual and corporate memberships available.

thewitnetwork.com

Christine Bongard, Founder

Corinne Sharp, Founder

13 THINGS MENTALLY STRONG WOMEN DON'T DO

1	They don't compare themselves to other people.
2	They don't insist on perfection.
3	They don't see vulnerability as a weakness.
4	They don't let self-doubt stop them from reaching their goals.
5	They don't overthink everything.
6	They don't avoid tough challenges.
7	They don't fear breaking the rules.
8	They don't put others down to lift themselves up.
9	They don't allow others to limit their potential.
10	They don't blame themselves when something goes wrong.
11	They don't stay silent.
12	They don't feel bad about reinventing themselves.
13	They don't fear owning their success.

Figure 1. From *13 Things Mentally Strong Women Don't Do*, forbes.com, 2019

We need our girls to recognize their mental acuity, capacity and ability at a young age. We need them to embrace how smart they are. We need them to recognize their incredible abilities early as they develop and hold on to that knowledge. We need them to fine-tune their abilities and develop them over their lifetime. We need them to have confidence in their abilities and know that their attributes are noteworthy and worth celebrating.

We have seen girls say "I'm not smart" or "I'm so stupid" at young ages when they want to fit in or not appear to be smart. They don't want their brilliance to stand out, and we need to change that. We need them to embrace how smart they are. A recent study indicated girls as early as six were opting out of Math and Science[3]. First, in most geographies that is grade one. They are just starting their school life. Second, they know how to count and we know they have yet to take a Science course, so this is troubling. If they are feeling this so early in their school life, imagine where this might take them by the time they start university. If these sentiments settle in, how can they ever lift themselves out? And most troubling, our beliefs become our reality. They become actions. They become self-fulfilling prophecies.

We often hear women have an incredible ability to take in a lot of information and retain it. We have a virtual filing cabinet in our heads. We have strong memories. We are known for holding a million details: birthdays, clothing sizes, school projects, homework, community matters, religious events . . . you name it, we remember it. We are the organizers of our lives, the CEOs of our worlds. Is this something that starts in childhood, or is this something that develops as we grow up and make decisions about our lives? Do girls have this ability and are they just not aware of it? What if we tapped into this gift at a young age?

At the age of two or three, I was told how smart I was. I was told I could do anything I put my mind to. I was told there was nothing I couldn't do if I set my mind to it, and to not let anyone ever tell me I couldn't do something. At that age I had no real sense of the power of those words, but they certainly created a foundation for the rest of my life. I was never diminished, put down, felt to be less than anyone else. In fact, I was taught the very opposite. The messages I heard in my home were quite positive and reinforced a strong belief in myself and my mental abilities. In fact, I was told my brain was one of the most important parts of me. When my great-uncle used to come home from Winnipeg and ask me "Who's the prettiest girl in Montreal?" my grandmother quickly said, "Don't put that foolishness in her head; she's smart." I never appreciated the significance of that comment until decades later. I was the eldest child in my family, and school was the most important thing in my life. My parents felt strongly about education and learning. I had other interests like tap and ballet, but school always came first. Our assignments and homework took precedence over play time, hobbies and time spent with friends. We could not watch TV until our schoolwork was done. Every day I came home from school I had my homework to do and was expected to help my siblings with their homework too. My parents struggled with the "new Math" so it was my job to teach my siblings and ensure they understood the concepts. I did not have an above average IQ or intellectual abilities that were out of the ordinary—at least I certainly did not have any evidence of that—but I loved to read, I liked to solve problems, I liked to learn and study and do well in school, so all of those efforts paid off with good grades. I simply loved everything about school. I felt at home in the classroom and my brain took in information at a capacity that I can now genuinely appreciate. My curiosity and thirst for learning fuelled me. I developed thinking skills and discipline in working through things I did not understand. If I didn't know or understand something, I asked questions. I took pride in solving things. I took pride in studying and getting good grades. My parents were proud. I think these skills I learned when I was young continued to develop over the rest of my life and they have served me well. As I got older, I discovered my brain worked differently at different times of the day. My brain and my ability to think and process information is best in the early part of the day when my thinking power is at its best. I discovered I was able to tackle and solve difficult problems during the morning hours, so I reserved that time for challenging tasks. That knowledge was an important life hack for me.

Young children, both girls and boys, are proud when they learn new things and eager to show others what they have learned. Children demonstrate this strength, or intellectual capacity, when they learn to tie their own shoes or make their own breakfast, and they are excited to apply this capability on a frequent basis.

That intellectual capacity does not go away, but we can see children hide this strength. We also see children afraid to "try and fail" enough times to learn a new skill, especially in front of a peer group as they enter their teenage years, or even in front of family members if they haven't been supported along the way.

Our words can impact children in a positive or negative manner. Think of the difference in asking in a disappointed tone why they didn't achieve the goal versus a positive tone encouraging them when they almost tackled it.

Many people describe themselves as lifelong learners. This has been described as the ongoing, voluntary and self-motivated pursuit of knowledge. I identify with this in my personal and professional life. It's fun to try new skills and often exhilarating when I achieve them.

I've always enjoyed creating something with words, whether it be through work presentations, personal blogging or writing books. When I was ten, I asked for a typewriter for a present because I thought it would be ideal to be able to get my thoughts on paper faster. It takes thought and time and effort to get better at a new skill. One of the best pieces of advice I've received, about writing and other learning, is to just start and be kind to yourself. Taking the first step is often the most challenging. Once you've done that, you can begin tracking your progress instead of stewing about whether or not you want to try to acquire this new skill. Be kind to yourself since you will not be great at the beginning, and it will take time and effort to make progress. If you already knew how to do it, it wouldn't be called learning.

It's difficult sometimes to put yourself out there in front of people at school, work or in your personal life. Taking that chance and being vulnerable is necessary to receive lessons, feedback, tips and techniques from others who know more than we do.

I enjoyed writing as a child. I enjoyed writing papers for school and authoring articles for the yearbook. I drafted poems at home for my own pleasure and journaled off and on for several years. As I wrote, I could feel the strength in learning to express myself and finding different venues to apply what I was learning. I did not actively seek out opportunities to speak in public, but I did seek opportunities to write. Writing gave me the chance to gather my thoughts, to edit when required, and to proofread or ask others to proofread the material before sharing it broadly.

I was proud of my ability to articulate points of view, and stories, and daily life events. As I started my career, I sought people who could help me learn more including being succinct, selling the point when required, and understanding the difference between when the message needed to be emotional versus strictly factual. I looked for opportunities to contribute through presentations, business cases and annual business plans. There are many times writing is required in our work lives, and I enjoyed being known as someone who would not only be able to create the right message but also to ensure the details were accurate and complete.

Writing at work was rewarding, but I also wanted to find avenues to write that were more personal. I started to write publicly in 2010 with an anonymous blog. This felt safer, to write what I wanted and not worry about reactions that could make me consider stopping. I started with a lifestyle blog www.workingtechmom.wordpress.com and was careful not to provide personal details. In this blog, my kids were simply called Big One and Little One. In 2024, I am still blogging at www.kelleyirwin.com/blog where the details are more personal.

My confidence as a writer has grown, and my ability to handle feedback has also grown. I am sharing my personal and professional journey through life to support others who may be having similar challenges and experiences.

Girls and boys can build these strengths and use them. It's important for parents and teachers and everyone influencing young people to encourage them to recognize their skills, learn new things, practise so they understand their own strengths and to never minimize them.

Let's hear from the girls

Meet Samaira, age 10

She is interested.

Samaira became interested in technology at the age of five. Her dad exposed her to tech when he showed her a programming language and explained how the code executed to make things happen. Her small stature and soft voice are at odds with her strength of conviction and her apparent intellectual strengths.

Samaira is interested in technology like robots that people can use without keyboards.

Her mom is a Homemaker and has worked in Creative Content writing and Management. Her dad works as a Technology Consultant.

She feels the power of tech.

She describes how she could build a robot to do the things she wants it to do. She doesn't want to buy a robot. She explains that kits should be made available to children so they can create a robot themselves and learn how to make it unique. She is confident we can learn to use the robot to solve real world problems.

She is taking action.

Samaira talked about her dad exposing her to technology, her teacher encouraging girls to participate, and organizations that introduce girls to the broader world of technology. She named organizations including Girls Who Game and Girls Who Code. She has a Raspberry PI kit, and a Talk Bot. She thinks about the oceans and how we can use robots to clean up the plastic.

When asked what she wants to do with her knowledge, learnings and experience with technology, Samaira replies without hesitation, "Change the world by solving big problems."

let's hear from the girls

Meet Scarlett, age 16

She is interested.

Scarlett became interested in technology at the age of 10 when she was exposed to robotics. She enjoyed the teamwork aspect with everyone working to solve a common problem. She demonstrates her strength in learning new skills and standing up for herself.

Scarlett is concerned there is a cultural issue in tech that occurs when you aren't around people you can relate to. In her Advanced Placement Computer Science (APCS) course, there are only three girls. Girls can be underestimated when only the boys are called on to solve the problem. We need to have more diverse teachers and guide girls on what they can do when a teacher tells them no. She describes one example where she felt the teacher could have just said she wasn't good enough "yet", and could have tried to teach her in the way that she could learn.

She is also interested in fashion and using technology to help create designs. She is an athlete, is on the Lacrosse team, and likes to draw and write poetry.

Her mom is a Lawyer who works on contracts for technology software and services. Her dad is a Chef.

She feels the power of tech.

She enjoys tech, the way everything comes together and how you can be creative with your own ideas or working with other people.

She has been inspired by Anouk Wipprecht[4], a Dutch fashion designer, who combines fashion and tech in her innovative designs. This is an emerging field of fashion tech and allows people to combine these two interests and talents.

She also highlights Karlie Kloss[5] who launched the program Kode with Klossy. This is an online program focused on creating learning experiences and opportunities for young women and gender expansive youth. Their goals include building a more equitable culture in technology and encouraging girls to pursue their passions in a tech-driven world.

She is taking action.

Scarlett participates in her middle school coding club. They work together to build projects using technology. She believes we need to create safe and welcoming environments for girls to learn.

She attended three summer camps to learn app development, web development and site making. She also worked as a summer intern on a Data Science team.

When asked what she wants to do with her knowledge, learnings and experience with technology, Scarlett quickly replies that we can solve any problem. We should focus on environmental issues as tech is a great way to address them. We should concentrate on ideas that will have lasting impact. The possibilities of tech are endless. We can advance and solve problems and, at the same time, open doors and bring people together.

Tips and Techniques

Girls

1. Celebrate your successes.
2. Speak up when you have information and knowledge to share.
3. Form opinions about what is happening in your life.
4. Learn from the things that go wrong.
5. Recognize nobody is perfect and do your best.

Parents and Teachers

1. Identify and name their personal strengths and talents.
2. Encourage them to try new things and support their progress.
3. Encourage participation in school and community activities.
4. Provide opportunities to learn new skills in their daily lives.
5. Encourage girls to compete against their previous attempts and goals, not against others.

Moving forward

We have an opportunity to recognize the intellectual strengths in girls from the time they are young and use our words and tone of voice to encourage their continued learning.

Learning and demonstrating new skills will open the door to many opportunities for impactful relationships, ongoing education, jobs and the pleasure of their accomplishments.

STRENGTHS OF FEMALE LEADERS

Aspect	Strength or Opportunity.
Perception	*Panoramic view.*
Dedication	*Pro-social engagement.*
Power	*Able to share power.*
Success	*Shares successes.*
Networking	*Connects projects and people.*

Figure 2. From Martina Kessler, ecstep.com/female-strengths, 2014

KNOW HER NAME

Ada Lovelace

COMPUTER PROGRAMMER

Ada Lovelace made her mark on the world in the mid-1800s, during a time when it was not only highly uncommon for a woman to learn STEM disciplines, but computers as we know them today didn't even exist yet.

Lovelace was fascinated with the brain and by other Science and Technology disciplines. In 1833, she met a man named Charles Babbage, who had created an early computing machine called the Analytical Engine. Lovelace translated one of Babbage's Spanish lectures into English and added notes. In her notes, she included an algorithm that allowed Babbage's engine to compute Bernoulli numbers and, as it turns out, this was the first time a computer algorithm had ever been published.

Because of her published algorithm, Lovelace is often considered to be the first ever computer programmer.[6]

KNOW HER NAME

Radia Perlman
MOTHER OF THE INTERNET

Radia Perlman attended the Massachusetts Institute of Technology in the late 1960s and early 1970s. During this time, few women were in these STEM programs.

Perlman's work has made a significant impact on the tech field — particularly on how networks move data and organize themselves. Her most notable creation was the spanning tree protocol (STP); a set of rules for network design that helped improve the internet. Others have expanded on the technology since Perlman invented it, but it was her creation that paved the way for the modern, ultrafast networks we enjoy today.[7]

She was inducted into the 'Internet Hall of Fame' in 2014 and into the 'National Inventors Hall of Fame' in 2016.

Courage

Courage—the quality of spirit that enables a person to face difficulty or danger.

Girls and boys can demonstrate courage in the face of adversity, achieving successes they believed were beyond their reach. It's important to recognize these moments, and congratulate and encourage them to continue striving beyond their comfort zone. We can see courage in tech by simply sharing an idea, identifying a solution and taking a risk. This includes assessment, analysis and often trial and error.

Assessment is the cornerstone. What is the problem you are trying to solve? What is the opportunity you are trying to create? Do you see a current product or process that could be enhanced? Do you see a need and no product to meet it? Assessment can be quietly done alone, or loudly done with others. Thinking, talking and writing on paper, whiteboards or computers. Clarifying what you know and don't know, and boiling it down to an idea that feels exciting and maybe even complicated. Once you have the assessment of the current state, you're ready to state your goal. Communicating this goal takes courage as you are committing your ideas to actions.

Analysis begins once you have a clear goal. Knowing what you want to achieve allows free thinking for how to get there. Pondering a myriad of ideas can open your mind to the possibilities. Think of options that could be easy or hard to create, that might meet all the goals you stated or a portion of them for the first step. This analysis can start broadly and then be narrowed down, or start small and evolve and grow. There is no correct path to analysis. The approach should be enjoyable and fluid. Think of impossible solutions, and then challenge yourself on why they seem impossible.

Trial and error involves testing out your ideas. This can feel exhilarating for some people and terrifying for others. Sharing your analysis and ideas for solutions with others takes courage. You can start with one person and sharing one idea. Identify a person who is on your side, someone who wants you to succeed. They might not know anything about the topic, but sharing your ideas helps you formulate your thoughts and identify gaps in your analysis or your solution options. Take that step. Share your idea. Listen to the feedback. And continue your analysis and testing steps.

We have a responsibility as we raise girls to encourage their individual spirit to navigate adversity and face the difficulties in their lives when they reach beyond their comfort zone.

DID YOU KNOW?

Girls Who Code is changing the game. They are on a mission to close the gender gap in tech. They have built the largest pipeline of women and nonbinary computer scientists in the world and, as of 2022, have officially served 580,000 students, including 185,000 college-aged alumni.

Reshma Saujani is the founder of Girls Who Code. She is also a lawyer, politician and a civil servant. Through Girls Who Code, she aims to increase the number of women in Computer Science and close the gender employment difference in that field.

Saujani is the author of Lead the Way, Girls Who Code: Learn to Code and Change the World and Brave, Not Perfect: Fear Less, Fail More, and Live Bolder.[8]

girlswhocode.com
Reshma Saujani, Founder
Tarika Barrett, CEO

6 THINGS CONFIDENT AND COURAGEOUS PEOPLE DON'T DO

1	*They don't try to please everyone all the time.*
2	*They don't worry about things that are out of their control.*
3	*They don't avoid new and challenging opportunities.*
4	*They don't get stuck on self-pity.*
5	*They don't spend time with negative people.*
6	*They don't need the approval of others.*

Figure 3. From The 6 things confident and courageous people don't do, lifecoach-directory.org.uk, 2017

Courage is a muscle. It is important for our young girls to be courageous. The more they use this muscle the stronger it will become, and they will need it to be strong. They must recognize they will have to stand up for themselves; they will have to speak up when something inappropriate is said. In school they will be faced with many different scenarios and situations that will require courage. We know girls want to belong, they want connection with others, and at times they may minimize themselves to fit in. It will take courage to navigate those situations. If they are not encouraged to be brave, they might not realize it is ok to be so.

When girls are toddlers, they are as brave as can be. They jump off the sofa, they climb the stairs as we parents bite our nails worrying they will fall. They run down the street, try to climb the monkey bars at the park, and try to get out of their cribs. They are curious and will do and try anything. That's bravery. But we parents and adults start chipping away at that. We start with messages like "don't climb so high, you will fall;" "be careful, you are going to get dirty." It is insidious and dangerous as we keep sending the wrong message. How can we expect them to navigate adolescence and teenage years when the pressures to be whatever someone else wants them to be is front and centre?

Today our girls are barraged with many different expectations, some quite harmful to their self-esteem and overall confidence. They are subjected to standards of appearance and societal norms that make it challenging for them to build their courage muscle. It is hard to step outside the crowd, to be comfortable in a developing body, to know what they think is ok, that they are perfect in every way and do not need to meet whatever artificial standards are being imposed upon them.

We all want to belong, we all want to have friends, we all want to feel loved, cared for and connected to those in our circle. It is imperative that our girls know that not everyone is going to like them or be like them, not everyone is going to agree with them, and that is okay. There are studies on the importance for girls to be confident in themselves as it impacts their outlook on how successful they can be in future career paths.[9]

The approval or expectations of others will always be there, for everyone, but the ability to separate from that, to think for yourself, to stand up for yourself, to believe in yourself, and to have the courage of your convictions can be incredibly challenging for girls. These pressures can have life-changing impact, unless our girls are taught at an early age to find a healthy balance between societal pressures and independent choices.

    I discovered my courage at four years old when I punched a boy twice my age in the nose. He called me the "n" word. I actually didn't know what the word meant, but he said it with such derision I knew he was calling me something awful. He yelled at me through the fence surrounding my nursery school playground. I went over to the fence and taunted him to come closer to me. When he got close enough, I grabbed his jacket and punched him in the nose. There was blood everywhere, and he was screaming. The nursery schoolteachers came running to see what had happened. I got in a lot of trouble for hitting him and cracking his nose. They said it was broken. They called my parents and asked them to come to the school and get me. That was the one and only time my parents were called to school because I was in trouble. My mother and father came and when my father asked the teacher why I had hit that boy and they told him what happened, my father said, "Then you are lucky that is all she broke." He took me from school and explained what the "n" word actually meant. He had to tell me it was a bad name that White people called Black people. When I asked him why, he told me, "They think we are less than them, that they are better than us." He then went on to explain and reinforce that they were wrong, and I was never to let anyone tell me otherwise.

We face difficulty from the time we are young. The difficulty, and the courage to face it, comes out in diverse ways.

When we think of the difficulty and not just the danger, we face this every day from the time we are tying our shoes, answering a question in our first grade of school, or making friends.

We learn to be courageous, to face difficulty and strengthen our abilities each time we face a challenge. Actions every day, every time, occur and we can point out these moments that girls have encountered to help them realize they are building this skill easier and more often than they realize.

There are also times when people close to the girls believe in them but may be holding them back in a certain moment or saying something that results in the girls holding themselves back.

The courage doesn't have to be physical—it can be emotional courage. Girls may feel the opposite of courage is being nervous as they are concerned what other people will say about them, or how they will feel if they get a wrong answer or try something new and they aren't the best. We can talk them through the situation allowing them to identify the worst thing they can imagine happening. We can then help them work through actions they can take to prepare for this potential scenario. We can also talk through how to react positively if the best outcome happens so they can appreciate the blessing of this turn of events.

I tried various activities as a child. I swam and found I was good at it. I played baseball and determined I was not. I explored Art and Music, Math competitions and Foreign Languages. I sang in the choir, took chemistry, and participated in Drama. Each experience was difficult at first. I learned with every attempt, not just what I was good at or interested in, that it became easier to be courageous over time.

I started Martial Arts as an adult with teenage children. I planned to train for a year to get the kids interested and then I would stop. I ended up committing to being a lifelong Martial Artist, achieving my black belt after ten years. I was scheduled to take my black belt test and realized the test was not just about the technical skill I had learned, but also about stamina, flexibility, and balance. There was another aspect that was making me nervous—thinking about the people who would be watching me, all my friends and family who knew I was testing, and how I would feel if I failed the test. It took a lot of courage to train for the ten years, but I realized I needed an extra dose of courage to put everything on the line for one day in front of all my classmates and all the black belts who were overseeing the test. I knew they were on my side and wanted me to succeed, but I was the only one who could do it. I needed the physical courage to make it through the seven hours of testing, and the emotional courage to do my best and live with the outcome.

Before the test, we were given words of encouragement that sounded like this, "You have trained, and you know what to do. Your head knows and your body knows. Today is about demonstrating that in front of people." It made me realize it is not like a test in school where you may or may not have studied the right sections, and you may or may not have understood them. I trained. I knew every move. And for those seven hours of black belt testing, I demonstrated it with everything I had in me.

In my Martial Arts class when a student is trying to achieve a challenging task, the other students create an environment of positive energy that is welcoming to the attempt. No matter what the outcome is, everyone claps and is supportive of the effort put forward to build the new skill. And when the student is successful, we feel it is an achievement of the whole class because we are all in it together. This day, like no other, showed me the power of summoning all my courage and facing the challenge.

Courage and motivation can come together in critical moments. If you think of the motivation you need to take the step, it will often get you over the moment and allow you to gain the courage. The motivation can be to stand up for others, to learn new skills or to simply have fun.

When we notice young girls being courageous, let's say it out loud—"You were so courageous, you have such great spirit, you should be proud!"

Let's hear from the girls

Meet Ella, age 10

She is interested.

Ella became interested in technology at the age of eight. She was exposed to tech on her dad's computer, and she played games including Snake. In the game, you feed the snake and it gets longer and longer. She then learned about online coding using the coding tool Scratch. It can be used for writing programs and sharing them. She built a program that took input of a person's name and turned it into colours. There is also a platform available for children as young as two to six years old called ScratchJr that provides coding blocks for easy use. Her natural curiosity and exploratory nature showed the courage of her actions as she tried a variety of technologies.

Ella thinks of Math as a method for problem-solving and treasure hunting. She is also active on social media, and her accounts are private with parental oversight. She likes being creative including writing code, making up stories and drawing.

Her mom is a trained Registered Nurse, Epidemiologist and Nurse Practitioner. Her dad is a trained Registered Nurse and Epidemiologist.

She feels the power of tech.

She describes how she started learning from code.org about different ways of coding. It is an education innovation nonprofit dedicated to the vision that every student in every school has the opportunity to learn Computer Science as part of their core K-12 education. They expand access to Computer Science in schools, with a focus on increasing participation by young women and students from other under-represented groups.

She uses computers in everyday classes to work on projects, search for information and write stories. She talked about Discord and WhatsApp– communication platforms that enable group chat as a foundation with other features.

She spoke of technology being used in positive (helping people) and negative (hacking/hurting people) ways. She is excited to use technology as an advantage to help people.

She sees the power of using technology leveraging 3D printing to make designs that are eye-catching and believes these types of features can make technology more appealing to girls. She discussed the need to also include links to additional material whenever engaging with the girls.

Her current school has mandatory computer classes starting in grade two, with equal distribution of girls and boys in the class. Starting in grade seven, computer classes are taught by a different teacher focused on Computer Science who identifies ways to engage the students. In high school, there are computer projects each week covering different topics to allow the students to explore the technology in areas that are of interest to them.

She is taking action.

Ella knows we need to make technology more appealing to young girls. Her ideas include making games that appeal to girls like Roblox. Pet simulator is a game on Roblox that allows you to collect, upgrade and trade various pets. She also wants to encourage girls to talk to their friends about the things they learn in tech to share the knowledge and the enjoyment.

When asked what she wants to do with her knowledge, learnings and experience with technology, Ella says she is interested in becoming a scientist. She wants to help people and understands that technology can be used to explore and address the source of problems like global warming and climate change. These issues affect all people in many ways, and we can use technology in our scientific study to identify actions to take.

Let's hear from the girls

Meet Ruby, age 7

She is interested.

Ruby became interested in technology at the age of four. She played educational games on her parents iPad and learned letters and numbers. She then discovered Minecraft, Roblox and Toca Life. She received a tablet of her own at the age of seven and enjoys games where you can create your own characters. She plays Genshin Impact, an imaginative world of adventure, where players complete quests as they travel around the map, unlock points and battle enemies.

In addition to technology, Ruby enjoys drawing and is interested in Japanese anime. She is cautious by nature and likes to observe activities before jumping in. She displays her courage in trying new things and talked about trying kayaking, canoeing and skiing.

Her mom works at a Software company. Her dad is a Product Executive and a data geek at heart.

She feels the power of tech.

For Math class, they get to use computing technology and also have homework that is often online. They use websites with reading games and Boom Learning, an online learning site that gives immediate feedback if you have answered correctly or not.

They are not coding in her class yet as that is done in higher grades at her school.

She is taking action.

Ruby enjoys learning and experimenting with technology platforms including multiple tablets and how they work differently. She was excited when she recently received the first tablet that was her own.

She has thought about what she wants to do when she grows up. She loves to bake and cook with her mom. She plans to open her own bakery and have the chance to give away food for free.

When asked what she wants to do with her knowledge, learnings and experience with technology, Ruby says she wants to use it to help people learn.

Tips and Techniques

Girls

1. Be yourself—always.
2. Understand that not everyone will like you and that is okay.
3. Stretch yourself to try new things and know that it is okay to feel afraid.
4. Give yourself time to be fearless.
5. Ask for help when you need it.

Parents and Teachers

1. Encourage your daughters and students to take risks.
2. Highlight the courage of other young girls.
3. Share your personal challenges and successes of being courageous.
4. Tell stories to expose them to role models of courage.
5. Talk to the girls about doing their best, not striving for perfection.

Moving forward

We have an opportunity to recognize the courage in young girls as they face challenges large and small. The danger we perceive can be a negative reaction or retaliation, verbal or physical, in response to our actions. By pointing out the danger they faced and the ability they had to overcome it, they will start to recognize they are braver than they think.

Facing difficulty or danger and achieving a positive outcome will prepare them for the larger challenges that will come later in life. The girls will start to understand they can assess the challenge, make appropriate decisions on how to tackle it, and be proud of their actions.

10 HABITS OF COURAGEOUS WOMEN

1	*They own their fears instead of waiting for bravery.*
2	*They stand up for what they believe in, even if it's unpopular.*
3	*They are strong and resilient.*
4	*They're not afraid to take chances and try new things.*
5	*They persevere even in the face of adversity.*
6	*They lead by example.*
7	*They have purpose.*
8	*They're compassionate and kind, even to strangers.*
9	*They practice calmness and self-care.*
10	*They support other women in leadership.*

Figure 4. From The 10 Habits of Courageous Women, https://soniamcdonald.com.au, 2022

KNOW HER NAME

Grace Hopper
COMPUTER PROGRAMMER

Grace Hopper joined the United States (U.S.) Navy during World War II and was assigned to program the Mark I computer. She continued to work in computing after the war, leading the team that created the first computer language compiler, which led to the popular COBOL language. She resumed active naval service at the age of 60, becoming a rear admiral before retiring in 1986.

Hopper's legacy includes encouraging young people to learn how to program. The Grace Hopper Celebration of Women in Computing Conference is a technical conference that encourages women to become part of the world of computing, while the Association for Computing Machinery offers a Grace Murray Hopper Award.

In 2016, Hopper was posthumously honored with the Presidential Medal of Freedom by U.S. President Barack Obama.[10]

KNOW HER NAME

Mary G Ross
ENGINEER

Mary G. Ross was part of the original engineering team at Lockheed's Missile Systems Division, where she worked on defense systems and contributed to space exploration related to the Apollo program and the Polaris re-entry vehicle. She is known as the first Native American woman engineer.[11]

As a key member of the Los Angeles chapter of the Society of Women Engineers, she is known for her commitment to improving educational opportunities for women and for all Native people, especially in the field of engineering.

She was featured on a commemorative U.S. dollar coin that celebrates the contributions made by Native people to the U.S. space program in 2019.[12]

Independence

*Independence—relying only on oneself
or one's own abilities, judgement.*

Girls and boys have the ability to demonstrate independence in their lives. Independence shows up as words and actions, in thoughts, in decisions, and in the ability to accept or reject ideas and plans that other people present.

Learning this skill is important for technical innovation. Being independent in thought means you can listen to an idea and assess it. You may want to accept the idea, reject the idea, or build on the idea if there are parts of it that you agree with.

Independence is important as you understand what you believe, what you want to pursue, what you want to achieve. This skill allows you to listen and think, to assess what you experience and evolve in your thinking with new facts and ideas.

Girls are often extremely independent when they are young. They do not hesitate to play by themselves; they are not afraid to be alone. They do various things on their own, whether it's creative activities, building something, talking to their imaginary playmates, reading a book, watching a show, playing on their devices, or exploring something of interest. We especially see this in toddlers who want to do everything by themselves, without any help from their parents or siblings. It is interesting to consider when that changed. When do they start putting others' needs before their own, and when does it happen that they will sacrifice their own desires and wishes to please other people? These are all important questions, character traits and attributes we must understand and combat. We want them to be kind, compassionate human beings, but we don't want them to do that at their own expense. It is not their job to take care of the world.

We have a responsibility as we raise girls to communicate that putting themselves first is not a selfish act.

DID YOU KNOW?

Girls in Tech (GIT) is a global nonprofit organization dedicated to eliminating the gender gap in tech with more than 130,000 members in 35 global chapters around the world. The focus is on engagement, education and empowerment of girls and women who are passionate about technology and are committed to building the diverse and inclusive tech workforce the world needs. They aim to see every person accepted, confident and valued in tech — just as they are. So, when you're asked how you fit into the industry, they want you to boldly say: "As I Am."

There are chapters in Armenia, Australia, USA, Dominican Republic, Ecuador, Finland, Germany, Greece, Indonesia, Jordan, Korea, Malasia, England, Macau, Mexico, Canada, South Asia, New Zealand, Israel, Poland, Singapore, South Africa, Spain, Switzerland, Taiwan, Tanzania, Uruguay and Vietnam.

girlsintech.org

Adriana Gascoigne, Founder

12 THINGS STRONG, INDEPENDENT GIRLS DON'T DO

1	They don't neglect their careers.
2	They don't fail to handle their own situations.
3	They don't overreact to bad situations or mistakes.
4	They don't rise to the bait of haters.
5	They don't stop learning.
6	They don't act on first impulse.
7	They don't let other people affect their confidence.
8	They don't neglect their physical needs.
9	They don't have unrealistic expectations.
10	They don't stay in toxic relationships.
11	They don't let someone else dictate their relationships.
12	They don't lose control of their lives.

Figure 5. From 12 Things Strong, Independent Girls Don't Do, LifeHack.org, 2017

Debra

Girls are wonderful. Girls are smart. Girls are resourceful. Girls are committed. Girls are independent. So, what happens to change this? What starts to chip away at that fierce independence? When do they start doubting themselves and shutting that down, or become needy, or seek approval for their thoughts, beliefs and actions? When was the last time you observed a two-year-old girl, fiercely independent, following her own path, stating what she wants quite openly and telling you no—often? She is determined to do whatever she has in her head, and she is not letting anyone get in the way; not you, not me and certainly not her parents or any other adult. Independence is essential for girls. The more independent they are, the harder it is for them to be influenced, challenged or even derailed. Independence means freedom. Freedom to think, freedom to choose, freedom to make their own decisions for their life. Independent girls are unstoppable. The more independent they are, the less likely they can be taken down a path that doesn't serve them well. Their independence gives them control over their thoughts, their body, their actions, their lives. Independence is essential.

Of course, they need to get to adulthood without their independence being totally curtailed, diminished or destroyed.

They emerge from childhood having spent two decades of living, navigating whatever life throws their way. All adults in their world play a pivotal role in fostering and encouraging their independence, knowing full well they will need to exist in a social structure that might think otherwise. Independence is important for so many reasons. They need to think for themselves, have their own ideas and thoughts. Being able to think and do for themselves, without depending on others, can be a beacon for their entire lives. They can decide who they want to be with, what they want to do, how much they want to commit to, how to respond to whatever situation they face, how much risk they want to take and how they want to live their lives.

Independence builds confidence, and that is important. The more confident they are in who they are and what they stand for, the more capable they are to make decisions at every stage of their life. We live in a society where influence is everywhere. Young girls are susceptible to the power of social media and its reach. Without a strong sense of character and independence, they can be easily influenced by destructive forces. Independence can protect them from the power of praise or criticism, both can have a detrimental impact on girls.

Young girls are highly independent, even more so than boys.[13]

We know this is not an easy path as the long-standing stereotypes are deep-seated in cultural and religious doctrines, and rising above these could mean stepping away from their families and the people they love.

As a young girl I had a mind of my own. I was not easily swayed or influenced by others. I was very opinionated and had a lot to say about everything. That came from my upbringing and being told there is right, and there is wrong, and do not let anyone make me do something I didn't want to do. I needed to think for myself, always. I had to make good decisions, do the right thing, always. It's not about being perfect but about being principled and to not bring shame on the family by getting in trouble. We were discouraged from making friends who did not share our values. The standards were high, and we were expected to meet them.

As a girl you do not realize the importance of these messages. I wonder now if my parents were so clear in these directives because they realized that Black children and Black girls could be so easily taken advantage of, or hurt, or compromised in some way. I wish I could ask those questions of my parents. Why were those messages so important? I understand it as a parent now because I've told my children the same message even though they have a Black mother, a White father, and they are White-presenting. So colour had nothing to do with the messages I delivered to them, but these messages were incredibly important. I wanted them to think for themselves, to not be blindly obedient to me, their father, or any adult. This could be because there were increasingly more incidents of children being harmed by adults, and I wanted them to know they could say no, they could determine for themselves what was right in any situation.

When I was growing up you could easily go down the wrong path. There were forces in the community that meant you harm. There were kids in school who would try to convince you to try things like drugs or alcohol.

I paid it forward with these messages in raising my children. I remember telling my 7-year-old son that if I ever got called to school for troublesome behaviour, he needed to tell me one thing—"that he was leading the pack". He questioned me to understand what I meant. My response was "If you are going to do something wrong, I need to know that your brain was engaged, you thought about what you were going to do, you might have known it was wrong and you did it anyway." I didn't want to hear him tell me he didn't know why he did it, or even worse he was following little Johnny who was the leader of the pack. I didn't want to hear he was simply following someone.

I was the oldest child, so I had a lot of responsibility for myself and my siblings. I had that responsibility at a young age. I also was focused on the things that mattered to my family, doing well in school and being part of our church family and the Black community overall. My independence was encouraged and nurtured. I took initiative and was comfortable in my own abilities. If I didn't know something I could learn it. I never doubted it. I never said I cannot do that. I didn't think I needed to ask permission to do something. I just did it. That has served me well my entire life. There will be many people who will say no to you. Have your own mind, your own thoughts and opinions and say yes to yourself.

Independence can give you a feeling of power.

I remember feeling independent at an early age and felt confident in that independence. I also realized that sometimes that independence was hard on other people, including my parents.

Kelley

It's important for us to talk to girls about how independence can be demonstrated and how to deal with the potential fallout that could occur. I'm not suggesting they should shy away from their independence or pretend they don't have it, but to be prepared for the potential reactions that could range from merely surprise to resistance.

Relying on your own ability and judgement doesn't mean you won't take input from others. You can ask people for information and opinions and gain insights from your own experience. You can read and ask questions and then trust your judgement based on all those inputs. It's important to be comfortable with your decisions without seeking approval for every position you take or decision you make.

As a child, I was happy to dress myself and make my own breakfast. When I was six, we had a toaster on a small shelf beside the table instead of on the counter. This meant it was at a level I could reach, and I made toast for my breakfast each day before school. I was proud of this and still remember the little shelf, the pop of the toast coming out and the butter melting on the bread.

I felt independent in school when I picked my classes. There were even times I selected a class that didn't interest my friends. I also knew I would feel okay in the class without any of my friends beside me, and our friendship wouldn't suffer. I trusted my judgement to follow my own path. The bonus was making new friends along the way.

I felt my independence at home when I sang to myself at night. My brothers shared a room, and I had my own room. I could hear them talking and I didn't have anyone in my room to talk to, so I sang to myself before sleeping. My family has always enjoyed music, so this was just an extension of that enjoyment. I'm not sure my brothers appreciated my solos as much as I did.

My parents provided support and guardrails of expectations. I knew I could take chances doing things alone, exploring activities, friends, classes and job opportunities. I was comfortable making these decisions knowing I could ask for help or advice if I found myself in a situation I couldn't manage.

I was talking to my mom recently about how families give each other advice. She commented that I never got irritated when a family member or friend provided unsolicited advice to me, even if the topic was one that I had more experience with than the person offering the advice. She asked how I was able to do that. I smiled and replied, "I just filter." We can listen to advice without taking it. We can also take a nugget from the advice if it is relevant without following the advice completely. This is what independent people do. Trust themselves, be open to new ideas and evolve their thinking when new facts are presented.

Independence can mean there are times when you may spend time alone or pursue an activity by yourself. It's important to follow your own interests even when this may mean taking a class, reading a book or participating in a social activity that your friends are not interested in.

Independence should be seen as a positive trait.

Let's hear from the girls

Meet Lydia, age 12

She is interested.

Lydia became interested in technology at the age of four. She made the connection to tech as she used the remote controller to change the channels on the television. Clicking the buttons made something happen and she experienced that sense of independence. At the age of six, she figured out the connection and became more interested in technology and computers.

Lydia likes soccer, musical theatre, cooking and using technology to "fly through space". She also writes in her spare time and is currently writing a mystery novel, "The Last Rule".

Her mom is a Social Worker working in the field of long term care and was the mentor for the robotics team for both of her children. Her dad is a Mold Maker at a tool and die factory.

She feels the power of tech.

As her mom led a robotics team, she was excited to see how the robots were created and how they moved. She started to build mini robots with Lego robotics kits, including a model of the Rover robot that was used on the moon. She has used an iPad in the past and is now focused on her PC and her mobile phone to create and consume technology.

She finds 3D modelling interesting and uses Animator to understand various shapes. She is exploring technical products like Roadblocks and games like Minecraft. She uses the Kleki App for online painting and a Virtual Reality (VR) headset for new experiences like flying through a space station.

She sees stereotypes impacting girls when they are led to think of using technology only for makeup and modelling, while boys are encouraged to use technology for mechanical advancements. She believes girls and boys working together to design and use the technology will bring a broader point of view. She also knows that teachers working step by step with students can expose them to all the options and break down the stereotypes.

Her school has many different computer classes. If the students are doing well in these courses, they can be sent to a higher grade where they will be exposed to additional technical capabilities and experiences.

She is taking action.

Lydia loves computer design and has taught herself to code. She recently started using Scratch at school and leverages Photoshop, designing characters, and adding stop motion animation as she plays Gacha Club. She has also designed various home interiors on her computer.

She leverages technology to interact with her cousin who lives far away.

She sees the positive contributions technology can make for our environment including wind turbines and reducing cars on the road. She also understands how video chat functions can be enhanced for improved long-distance communications.

When asked what she wants to do with her knowledge, learnings and experience with technology, Lydia quickly responds she wants to work on technical changes to prevent cyber bullying.

Let's hear from the girls

Meet Nakia, age 15

She is interested.

Nakia was exposed to technology at the age of two. She is the oldest of eight children and attended a Montessori school when she was young. Her parents found the school programs and instruction contributed to her advancement in Math and Sciences.

Nakia started experimenting with tech in the second grade with software like Angry Birds. Her school had technology programs in her elementary years of grades four through eight and she demonstrated her independence as she relied on her own abilities, and she grew these skills.

Her mom is a Nurse. Her dad is a Record Label Producer and a retired professional Football Player.

She feels the power of tech.

She currently attends a Science, Technology and Industry school (STI) that offers College Preparatory, Advanced Placement, Science, Technology, Industry, Business and General programs. Students apply to the school in their eighth grade. The school offers 16 different technology streams, including networking and cybersecurity.

She has learned about artificial language (AI) and machine learning (ML) with formal education in networking, cybersecurity and engineering concepts. She is also learning programming as part of the curriculum.

She feels that schools should not generalize or stereotype. She is positive there are many options available for girls and boys, and schools should educate all students on the many different sub-areas of technology. Her advice to girls is, "Don't give up, don't beat yourself up, keep trying, and research different areas of interest until you find what you like."

She spoke of programs like MESA at John Hopkins University that provide opportunity for girls. MESA (Math, Engineering, Science Achievement) is an exciting and engaging after-school program for students in grades 3–12 designed to spark their interest in STEM education and STEM careers. The program seeks to increase the number of Engineers, Scientists, Mathematicians and related professionals at technical and management levels, and encourage and assist minorities and females in achieving success in these fields.

She also mentioned Women Inspiring Girls, Women in STEM Excellence (WISE) and Skills USA as resources to help girls pursue their interests in technology and skilled trades.

She is taking action.

Nakia is planning ahead with a goal to attend Stanford, Georgia Tech or the University of California to continue her education.

When asked what she wants to do with her knowledge, learnings and experience with technology, Nakia states that she would change the way people view each other. She would make something that allows you to see how they think and feel completely. She thinks that understanding people we dislike or disagree with would change our perspective. She thinks the world would be a much better place if we treated each other with respect and kindness. If we took a moment to think and just try to understand each other, a lot of hate, negativity and violence would be eliminated from society.

Tips and Techniques

Girls

1. Think for yourself.
2. Challenge anyone who says you can't do something because you are a girl.
3. Be curious and have a positive mindset.
4. Try new things and enjoy learning.
5. Experiment and do things for yourself.

Parents and Teachers

1. Find examples to show them what boldness and independence look like.
2. Teach them how to make decisions as early as possible.
3. Challenge the stereotypes and expectations of girls.
4. Encourage self-reliance.
5. Teach them to solve problems by themselves, and to set goals.

Moving forward

We have an opportunity to change the trajectory in the life of a young girl when we foster independence early. Teaching them to rely on their own judgement and abilities means they will grow that independence in relation to their maturity and their increased knowledge and experience over the years.

Building that confidence in their own judgement will prepare them to ask questions and gather input, and then be comfortable with the decisions they make as they carefully consider all their personal knowledge, abilities and goals.

6 WAYS TO TEACH YOUR DAUGHTER INDEPENDENCE

1	Be her role model.
2	Coach from the sidelines.
3	Set high but attainable standards.
4	Strive for consistency.
5	Be on the same page.
6	Maintain your bond.

Figure 6. From 6 WAYS TO TEACH YOUR DAUGHTER INDEPENDENCE, thestartupsquad.com, 2019

KNOW HER NAME

Roshni Nadar Malhotra
CHAIR HCL TECHNOLOGIES

As Chair of HCL Technologies, one of the world's leading technology companies since 2020, Roshni Nadar Malhotra is the first female chairperson of a leading IT company listed in India.

She has been listed in Forbes' World's 100 Most Powerful Women for three years and was recognized by Horasis as the Indian Business leader of the Year 2019. She is an inspirational speaker and leads on numerous philanthropic initiatives. Roshni is a member of the Dean's Advisory Council at the MIT School of Engineering, USA, and is also a member of the Kellogg School of Management Executive Board for Asia. She has an MBA from Kellogg.[14]

KNOW HER NAME

Neha Parikh

CEO, FORTUNE 500 BOARD MEMBER

Neha Parikh held the position of CEO of Waze from 2021-2023. She previously led the hospitality brand Expedia Group, was the President of Hotwire and former VP of Hotels.com. Neha has deep experience growing, scaling and reinvigorating teams and businesses, and fiercely believes that real magic happens when you work on something you love with people who make you better. Neha was named as part of the inaugural 'Forbes CEO Next 50 Leaders Set to Revolutionize Business' and is a passionate advocate and speaker who is committed to helping others through her own journey. Neha also sits on the Board of Directors of Carvana, a leading e-commerce platform and Fortune 500 company. She holds a Bachelors of Business degree from The University of Texas at Austin and an MBA from the Kellogg School of Management at Northwestern University.[15]

Leadership

Leadership—the ability of an individual to influence and guide others.

Girls and boys have the ability to exhibit leadership. It's important to think about this skill, when and how to use it, and to increase the ability to lead over time. Leadership styles can include distinct types of influence over the outcome, including coaching, pacesetting, democratic, visionary, commanding and affiliative.

Coaching people involves leading by mentoring. Pacesetting includes leading by example as you show them the way. Democratic leadership is executed by encouraging participation. Visionary leadership is accomplished through inspiring people. Commanding is a form of leadership that requires establishing and communicating procedures. Affiliative leadership is completely focused on the people and relationships in an organization.[16]

It's important to learn distinctive styles and understand when to use them. People often have a preference and will default to that style at times. Leading people requires the ability to assess the situation, the person and the urgency and then engage in the style that will bring the best result.

We have a responsibility as we raise girls to help them identify several types of leadership and the occasions where each can be used effectively. As they develop their personal preference, we can help them understand when it is most effective and when it may need to be adjusted for positive impact.

DID YOU KNOW?

Leading Cyber Ladies is a global professional network of women in cybersecurity. They have established chapters in London and Europe, New York, Tel Aviv, Tokyo and Toronto.

Leading Cyber Ladies is a movement to get women of all kinds in cybersecurity together at meetups, to give talks and speak up about their work, get comfortable with public speaking and network with other such amazing ladies in a comfortable, professional, and friendly environment. They believe in creating a safe and friendly environment for women in the cybersecurity community, with the goal of achieving greater diversity, representation and equality in this vibrant field.

leadingcyberladies.com

Hila Meller, Founder

Keren Elazari, Co-Founder

Helen Oakley, Co-Founder

Bobbi Montgomery Heath, Co-Founder

Kana Shinoda, Co-Founder

5 THINGS GREAT LEADERS DON'T DO

1	They don't ignore criticism.
2	They don't let their emotions take control.
3	They don't avoid responsibility for their choices.
4	They don't break their commitments.
5	They never say never.

Figure 7. From 5 Things Great Leaders Don't Do, thefutureorganization.com, 2022

Debra

Witnessing leadership in children is a fascinating experience. And young children, girls in particular, automatically take command. Just watch a two-year girl with a group of her friends. She is organizing, directing, telling, advising and doesn't hesitate to make things happen. It is impressive to watch girls recognize what they like to do, get focused on what they want to achieve and come to know their talents and special skills. Some girls are born leaders who naturally take charge, and others develop these skills as they mature. We just don't label it that way. We label girls bossy, and we commend boys for taking initiative. We need to change our language and encourage skills that will serve girls well their entire life.

When we think of leadership a long list of attributes comes to mind—strong, resilient, trustworthy, principled, courageous, visionary, inspiring and motivating. If I had to choose only one word, it would be impact, the impact leaders have on others, whether they are coaching an athlete, training someone to play a musical instrument, leading a toy drive in their neighbourhood or raising funds selling chocolate bars or cookies door to door.

Leaders influence others. They influence by what they say and more importantly by what they do. They influence by showing others the way. They influence by showing they care. They influence by providing help or guidance. They influence by sharing their experiences that can help others navigate through difficult situations.

Leaders bring their skills, knowledge, experience, creativity and new ways of thinking to any situation. They collaborate and engage with others to get things done. They do not back away from a challenge. They persevere through whatever obstacles they have to face. They focus on the solution and not the problem. They listen to others, make decisions and move forward.

I had a very strong voice as a young girl. I was always running the show, gathering the neighbourhood kids in our backyard and telling them what we were going to do. I always had a plan or could figure out a plan. If there was a problem, I could find a solution. I was also responsible, and other parents trusted their younger children were being supervised when we played outside. My calling was to take that responsibility and make my parents proud. I was raised to achieve. It was important to my family so it was important to me.

We hear terms like "born leader." I was one of those people. I was strong-willed, strong-minded and didn't back down easily. Telling me what to do (unless you were my parents) was difficult for anyone to attempt. I did respect authority figures—teachers, parents, grandparents, aunts and uncles but I would not do something I didn't feel was right. I stood tall so I was principled in my position.

With a well-developed voice at a young age, I would step forward and protect other children when people tried to take advantage of them. I could address any situation and if I couldn't, I knew who could. I was fearless, taking a strong position and guiding others. This skill developed over the course of my schooling and my career. Others saw something in my ability to speak up and offer my thoughts, to take the initiative in handling tough situations and influencing others with my belief that there was a better way. Others believed in me from a young age and as I started my career, I was given the opportunity to use my skills and lead. I embraced it all. I still do.

Kelley Leadership can be portrayed at any age. The ability to influence or guide others doesn't mean you are a boss, or that you have all the answers or that you are in a business situation.

You can lead others in a charity drive, or a sports competition or even by influencing the decision a group is making. Leadership does not require you to be an official leader of a team.

We can help girls understand they have this opportunity in typical daily activities, and they can become stronger in their leadership skills as they live and grow.

If we think of young girls babysitting, or guiding other children to get on a bus or leading games in the playground, they are building that leadership skill. Leadership is not defined as the one person who is in charge; it is really a broad category of people who are influencing others.

We don't need to ask in group projects at school or at work, "Who is going to be the leader?" because this can and often will diminish the chances of the other people participating. Each person should bring their skills as leaders to create a better outcome in the end.

When I was in high school, I swam on a competitive swim

team in the summer and a different team in the winter. I enjoyed swimming and realized schools that had pools also competed in the State swim meets. My school didn't have a pool and therefore, didn't have a swim team. I worked with one of my friends who also swam and we produced a plan. What if we could convince the pool where we trained for community meets to allow us to use the facility for school practice and school meets? As we started to work the plan, we found that we didn't originally understand all the pieces that needed to fall into place. In addition to a pool, we would need a teacher to be our coach and sponsor. We would need approval from the principal and the school board for the team to be established. We would need permission slips and money for travel and would have to be registered in the state as a school in the division so we would be included in the schedules. We successfully completed all the approvals and logistics, we rallied a small team of swimmers and we were off and running.

One day, the local paper came and interviewed me. It was exhilarating! I was thrilled to compete on the school team in addition to the community teams for two years. We made it to the state championship, and we enjoyed ourselves along the way. None of the girls on our team were moving on to make swimming a career. We were simply happy to have this experience as a team, both in the water competing and outside the pool as we learned about organizing. This experience solidified in my mind how I could influence people and outcomes in a way that was beneficial for myself and others. This stuck with me throughout my schooling and my career. If you have the best interest of the group in mind to achieve a positive outcome, it is a powerful motivator to lead.

This feeling, of influencing a positive outcome, is a central theme in my life. My parents both demonstrated this leadership trait at home, in their jobs and in their community. They taught me to plan, lead, consider others, and guide people as I moved forward.

Leadership isn't about hierarchy. It is about the ability to guide and influence others. And girls should be encouraged to demonstrate this skill and applauded when they do it well.

Let's hear from the girls

Meet Shreya, age 15

She is interested.

Shreya became interested in technology at the age of 12 in middle school as she recognized real-world examples of leadership where technology was being used to solve issues.

Shreya moved from India to the United States at 10 years old, grade four. She enjoys the creative aspect of both Dance and Art. Her parents are involved in technology but not developing code.

Her mom is in Management. Her dad is involved in IT Procurement.

She feels the power of tech.

She got involved in a robotics team at school. Their challenge was to design an app or a prototype that would solve a real-world problem. Her team created a prototype for a user interface using Figma that would make exercise programs available for all children. This included fun games to motivate kids. She and her team became finalists in FIRST Robotics in the grade nine competition.

She also joined SPARK in grade nine. SPARK is a youth-led organization for hackathons using a business case methodology to create pitches and have them assessed by a panel of judges. She participated in creating a pitch to solve problems like how the theatre and media industries are plummeting.

Using her artistic talents, she worked with Youth Culture programs designing various illustrations to break the general gender and racial stereotypes of certain work fields with a goal to encourage diversity and inclusion.

She is taking action.

Shreya hopes to launch a "Learnathon" to help kids tackle big problems like sustainability and learn how technology like AI, VR, Block Chain can help solve business problems bringing the corporate, government and non-profit worlds together to change the world. She wants to use both a Science fair-type model and workshops with business professionals and experts to help.

When asked what she wants to do with her knowledge, learnings and experience with technology, Shreya describes using technology for medical innovation and research. "We need to leverage technology in ways to impact millions of people. We are able to automate administrative tasks to gain economies of scale and free up time for doctors to provide an interactive, wholesome and considerate patient experience."

let's hear from the girls

Meet Emma, age 8

She is interested.

Emma became interested in tech at the age of three. She started playing an online game, ABCya, and enjoyed learning new skills that increased in difficulty as you progressed through the grade levels. These games were focused on learning about Art, Math, Language and different cultures. She exhibits leadership as she talks about ways to influence others through her artistic creations.

Emma loves to draw and is quite good at it. She is interested in Japanese anime and using this in her artwork. Anime is a shortened form of the Japanese word animēshon or "animation". She is enrolled in a local Art school and is building a comic by continuing to add content at every lesson. The lesson each week increases in difficulty as they build on the previous lessons and build out the finished comic.

Her mom is a Strategic Business Advisor. Her dad is a Manager of IT Service Delivery.

She feels the power of tech.

She received a Nintendo and initially played the Animal Crossing game. In this game, players execute tasks to build their own home and decorate it while paying off the loan on the house. She built her house and then built the community around it and planted fruit trees. She selects the types of seeds to plant, and the colour of paint for her house. She enjoyed selecting the colours, purple and pink, for her house and decorating with an aquarium theme because she loves dolphins.

She is enrolled in Best Brains, a Mathematics program outside of school. She plans to eventually start coding.

She is taking action.

Emma is creative and artistic, and is combining her love of Arts with technology. She is exploring technology on her devices and using her skills to challenge herself. She wants to be a doctor so she can help people.

When asked what she wants to do with her knowledge, learnings and experience with technology, Emma talked about the ability to use technology in the creation of all genres of music. She may find herself using technology often in the music she loves.

Tips and Techniques

Girls

1. Find opportunities to step forward and lead.
2. Learn about other girls who are leading and learn from them.
3. Do not impose any boundaries on yourself.
4. Volunteer at school/church/community.
5. Use your strengths to define how you want to lead. Lead in your own way.

Parents and Teachers

1. Encourage taking charge and taking chances at school and in the community.
2. Share books and movies of women leading the way.
3. Be deliberate with your words as you message to girls.
4. Listen carefully to how girls talk about themselves and help them highlight their strengths.
5. Find resources for them—clubs, associations and groups where leadership is encouraged and developed.

Moving forward

We have an opportunity to celebrate leadership skills in young girls as they guide others through their words and actions. Highlighting the impact they have had on other people and on the outcome will raise their awareness of their own personal abilities.

Practicing leadership at an early age will develop the skill at a time they can receive feedback and adjust their personal style. This will prepare them for broader leadership opportunities in their future.

4 ACTIONS TO RAISE YOUR GIRL INTO A NATURAL LEADER

1	*There is no one alive who is youer than you.*
2	*It's not bragging if it's true.*
3	*She's not bossy, she's a leader.*
4	*Finding your own North Star.*

Figure 8. From Leadership In Young Girls, girlsthatcreate.com, 2021

KNOW HER NAME

Josephine Cheng
ENGINEER

As the Vice President of IBM Research—Almaden in San Jose, California, Josephine Cheng oversaw more than 400 scientists and engineers doing exploratory and applied research in various hardware, software and service areas, including nanotechnology, materials science, storage systems, data management, web technologies, workplace practices and user interfaces.

For almost three decades, Cheng has been a leader in relational database technology. She helped produce technologies like DB2 World-Wide Web and XML Extender for DB2.

Cheng has received numerous awards for her work, including Asian American Engineer of the Year in 2003.

She has been awarded 28 patents for her inventions.[17]

KNOW HER NAME

Dr. Adele Goldberg
COMPUTER SCIENTIST, ENTREPRENEUR, MOTHER AND EDUCATOR

Dr. Adele Goldberg is one of the inventors of the programming language Smalltalk-80. This language helped make computers accessible to the masses, turning displays of lines of text into a digital representation of something even our ancestors would recognize: a desktop.

She received the ACM Software Systems Award in 1987. She was included into Forbes' "Twenty Who Matter". She received PC Magazine's Lifetime Achievement Award in 1996. In 2010, she was admitted into the Women in Technology International (WITI) Hall of Fame.[18]

Assertiveness

Assertiveness—the ability to communicate your wants and needs authoritatively, while remaining fair and empathetic.

Girls and boys have the capability to develop assertiveness. Practicing and listening to feedback will result in the strength to demonstrate this skill that can benefit them in situations at school, work, home and in their community.

Being assertive includes believing in yourself, learning how to say no and being simple and direct.

When you value yourself, it is easier to communicate your needs. If you establish what is important to you, learning to say no opens you up for other opportunities that are of your own choosing. Being simple and direct reduces the chance of being misunderstood as you stand up for what matters to you and to what you can commit.

Learning this skill can be rewarding and reduce stress for you over time. As you learn to speak up, it will take you less time to resolve issues. You will spend less time stewing over situations that may be easier to solve than you thought. There are moments in everyday life you can practise this skill. You can also ask a buddy to be there with you as you practise the skill, so you don't feel alone. Ask them not to step in but stand beside you. You may be surprised how comfortable it feels to speak up for what you want by having a friendly face to turn to if you are at a temporary loss for words.

Start small as you practise being assertive. Start with things that are factual as these are less emotional and may not create as much anxiety. Speak up. You can do it gently, but don't just let it slide. Stay calm. Be respectful. Stay focused.

Being assertive is a positive trait when done calmly with respect and honesty.

We have a responsibility as we raise girls to help them put their wants and needs into words. We can then provide opportunities for them to practise stating those needs authoritatively and also practise saying no to requests that are not aligned with their values or may compete with the time and attention they need for current commitments.

DID YOU KNOW?

The **Economic and Social Commission for Asia and the Pacific (ESCAP)** is the most inclusive intergovernmental platform in the Asia-Pacific region. The Commission has headquarters in Bangkok, Thailand, and promotes cooperation among its 53 member states and nine associate members in pursuit of solutions to sustainable development challenges. ESCAP is one of the five regional commissions of the United Nations.

Getting more women into careers in technology starts with breaking down the gender stereotypes that prevent girls from studying STEM subjects. Through comprehensive changes to the way STEM subjects are taught, governments can foster girls' enthusiasm for technology, expanding the future digital workforce.

unescap.org
Ms. Armida Salsiah Alisjahbana, Executive Secretary of ESCAP

7 THINGS ASSERTIVE WOMEN DON'T DO

1	They don't disrespect the opinions of others.
2	They don't try to control the opinions or actions of others through force or manipulation.
3	They don't avoid conflict at any cost.
4	They don't feel out the situation and adjust their opinion to match others.
5	They don't give and give to others causing emotional and/or physical detriment to themselves.
6	They don't avoid conflict and allow their feelings to fester.
7	They don't over apologise, especially for things they didn't cause.

Figure 9.

Debra

There is a double standard for girls when it comes to being assertive. From a young age, girls assert themselves in all sorts of ways. They know what they want, they know how they want it, and they don't hesitate to express their opinions concerning what they want to do or read, or places they would like to go. We find it cute, even charming and bold when they are young. We smile. We are impressed. We encourage their own self-expression. And then something changes. We start using different language to describe their assertiveness. The first one that comes to mind is bossy, which has a negative connotation. So what happens?

Being able to assert yourself from a young age through your entire adult life is such a critical life skill. There comes a time when we start telling girls to quiet down, not speak so boldly, or change their tone. Of course, there are circumstances where they may need to speak more quietly—in church perhaps or a movie theatre. The locale might actually warrant a whisper. But those are not the circumstances I am referring to. In general, we start to shut them down with our own adult words or guidance. Gender stereotypes are in play. We expect girls to be soft, to be demure, to be polite, to be people pleasers. We have expectations of girls that we do not have of boys.

Being assertive is tied to confidence and our ability to express ourselves. It is tied to our own beliefs and to our belief in ourselves, our opinions, our perspectives and our point of view. We should be encouraging girls to express these beliefs openly and often. We need them to be confident in their voice that starts to develop at a young age. They have phenomenal social skills that equip them to determine what tone is appropriate or how they should frame something. Girls care and collaborate, so they have an innate ability to communicate extremely well. They have been trained from an early age to be empathetic and caring. They have been taught to listen. They have been taught to care about people's feelings, in some cases to their own detriment. They have been taught to be people-pleasers, so they pay attention to how other people feel.

We can teach them what assertive communications look like, but we must also teach them not everyone will appreciate it. We need to equip our girls, so they are not shocked or unprepared for others' reactions to them. They must be taught to not take things personally or to take others' emotions on themselves. They are well-positioned to develop their assertiveness skills. What we need to ensure is they don't lose themselves along the way. We need to encourage them, banter with them, discuss different topics and help them to be decisive. Practice with them when you are each voicing opposing views on a matter. Debrief and congratulate them when you are finished. Having opposing views is good. It is how problems are solved.

Assertiveness versus arrogance is something we need to discuss in great depth where girls are concerned. It is one of those traits that falls into a serious double standard. When girls are assertive, they are told they are bossy. When they speak their minds, they are told to tone it down. When they state what they want, they are domineering. No one wants them to be arrogant, not our boys or our girls. And most of us as adults do not appreciate dealing with arrogant people. We feel ill-equipped that they rub us the wrong way. Being assertive when you are a boy is considered bold and confident and those are positive terms, unless you are a female.

I am not sure when I first realized how assertive I could be. Because I had developed my voice at such a young age, assertiveness went hand in hand with it. But I know speaking with authority has served me well. It is also a bit of a double-edged sword as it can make others hesitate to challenge or question you. I am not always right, but I am assertive. When I took over my first big leadership role, I was just 30 years old. I had been a manager before, but this role was important. I sat at a leadership table with eight men, all White, all 10-20 years older, and they definitely looked at me like I didn't belong there. I had been promoted from a field level account system engineering to a leadership role at our head office. It was daunting for me to just walk into the room. And there was no question those men intentionally tried to intimidate me. I remember a meeting where I was offering my thoughts and ideas, and they were being dismissive—like what could I know about logistics. I got through the meeting, held my own and then walked into my VP's office, the bold leader who had hired me. He was determined to transform IT into a service-driven, customer-focused organization. That I could do. But it was going to be challenging if I hit a roadblock every step of the way.

So, as I stood in his office and tears of frustration ran down my face, and he quietly handed me a Kleenex box I told him directly, "I may be upset at this minute, please ignore the tears, I will get them under control, and you will never see them again. If that group of arrogant men, with antiquated ways of working think they are going to take me out or stop me from doing what you hired me to do, they better buckle up as they don't know who they are dealing with."

Being assertive should be seen as a positive trait. Let's encourage girls to feel this way and to be comfortable expressing themselves.

Girls have the ability to communicate their wants and needs from an early age. We should listen to them and be careful not to judge them for wanting to attain something that is aspirational.

Leadership, promise, courage, authority—these words are often used when describing boys who show assertiveness. Girls showing assertiveness are frequently described as bossy. This is a word that is seldom used to describe a boy so let's reshape our reaction to girls when we see them being strong and assertive and stating their goals.

Assertively stating your opinions and aspirations while being fair and empathetic is a challenging thing to do. If we celebrate girls when we see them doing this well, they will remember and build on this capability for their future. Wanting something and stating your intentions is not rude or disrespectful, even if someone else wants the same thing and only one person can achieve it. This could be the captain of a sports team, the manager of a work team, the lead in a play, the tech lead on a computer project or simply the person to go first in a lunch line.

I remember when I was young, going into a competition and

having someone advise me to stand back and allow others to go first. This is conditioning that often happens to girls. My Dad leaned down and whispered in my ear "Go get 'em Kel!" and that lesson stuck with me my whole life. He was telling me it was okay to want something and to go for it and not to intentionally let other people win.

When I was in high school, I played the clarinet and was in the marching band. I had played clarinet for many years and was good at it, but not great. I had a friend who also played, and she was first-chair clarinet in our concert band. I watched her play and the joy on her face was obvious. I did not have that feeling. I approached our band director with an idea to switch to the percussion section. He was not supportive of this move for many reasons, least of which was that I had no experience with any instrument in the percussion section.

I had a few conversations with him, stating that I needed a change, and I wanted him to help me make that happen. I was clear that continuing to play the clarinet in the marching band was not an outcome that would work for me. We came to an agreement that I would continue to play the clarinet in the concert band, and he would move me to the percussion section for the marching band. This resulted in my playing a bass drum.

The drum was BIG, and I hadn't thought about the challenge of being 5'3" and marching with it, but I was happy. I had stated my case and achieved the result I wanted. Each day when we practised and competed, I had to lay across the drum while the drumming coach hooked the straps to my back and lifted me up. I spent the next two years playing that bass drum, enjoying marching band much more than I had before. I also learned a lot about the drumming and percussion sections. Our band won state competitions and I like to think I played a key role with my steady hand on the drum playing the cadence that allowed everyone to keep in step on the field. I can still play the main street beat pattern we used for parades to this very day. That's joy!

We can be assertive and respectful at the same time. It's okay to ask for what we want and to stand up for ourselves.

Let's hear from the girls

Meet Sage, age 15

She is interested.

Sage became interested in technology at the age of 10. She was exposed to the programming language Scratch at school. Her mom got connected to the African Canadian Christian Network (ACCN)[19] when they came to Canada. This organization helps Black children have the opportunity to apply to independent schools in Ontario. She attends an "eco STEM" school which focuses on STEM, ecology and sustainability programs. They promote parental involvement by focusing on what is right for the individual child. This includes parents communicating to their children they can do what they want to do, and the children being assertive enough to communicate their interests.

Sage also likes music and acting in plays, Math and treasure hunting.

Her mom is a Co-ordinator of Black Initiatives at a major university with credentials in program and project management. Her dad is a Business Performance professional in the consumer-packaged goods (CPG) industry.

She feels the power of tech.

She started Computer Science courses this year. There is both a Computer program and a Robotics club at her school. She is also interested in psychology and learning about how the mind works, and why people do what they do.

She is involved in Performance Art, Singing and Theatre. Her interest in the Arts led her to "Soul Pepper"[20], a Performing Arts program where she learned to write plays and monologues. She plans to attend a school in the future that provides a variety of programs citing Julliard and the New York Conservatory as examples. She has visited both the University of Waterloo and University of Toronto in Canada.

She is taking action.

Sage believes we can help girls to be successful by starting early and making an emotional connection for them to technology. We need to talk to them by relating technology capabilities to the things they like to do and helping them create options that will provide a future path. Find ways to start small and get them interested and girls will respond.

When asked what she wants to do with her knowledge, learnings and experience with technology, Sage thinks of a career in law. If she could solve just one problem using the power of technology, it would be to reduce poverty.

let's hear from the girls

Meet Sydney, age 17

She is interested.

Sydney became interested in technology at the age of four. She enjoyed using the computers available to her, trying new things and playing games on the computer.

She participates in a variety of sports and has been a competitive swimmer for the past nine years. The 200-fly is her best stroke and she qualified for Nationals and Olympic trials last year. She talks about the technology that is integral to sports including the timing sensors in the starting blocks for swim meets. There are also integrated speakers so each competitor can clearly hear the announcement, "Take your mark; go!"

She is assertive in her thoughts and actions, communicating her wants and needs around her education and her future aspirations. She chose her university for next year based on both the ability to continue swimming competitively and the exposure of being in the centre of government to open opportunities for her career.

Her moms are an Executive Recruiter and a Technical Writer.

She feels the power of tech.

She recalls working on a research project in grade three. Her group used technology to search online information about the trillium flower and printing materials. She also used iMovie to create the presentation.

She uses technology every day for school including online tests and Google Docs to stay organized. They use the D2L Hub, an integrated platform that creates a single source for project assignments and school announcements. This minimizes the need for students and teachers to use paper.

Her school has Chromebooks that you can use, but she brings her laptop with her each day. She finds it a critical tool for her personal and school life to stay organized and informed, and to communicate with her teachers. She can remember when her first laptop broke down and felt the impact of not having access to technology.

She is taking action.

Sydney believes we need to provide more opportunities for young students to participate in tech programs and make these courses mandatory similar to Math, English and Art. At her school, computer courses were elective and not part of the core program. Making these programs more prominent at a younger age would help them stick.

When asked what she wants to do with her knowledge, learnings and experience with technology, Sydney believes we can make a difference in addressing inequality. Social media can be used to profile communities that experience bias and discrimination based on gender, race, and sexual orientation. Raising awareness of how people are treated, particularly in the LGBTQ+ community, and the challenges they experience can provide a better understanding and encourage positive change.

She wants to work in international government when she graduates and is interested in being a diplomat, working at embassies and helping with government relations.

Tips and Techniques

Girls

1. Form your own opinion; find your unique perspective.
2. Share your thoughts often and openly and ask others what they think of your ideas.
3. Refrain from saying "whatever you want". Make a choice and be comfortable stating it.
4. Find places you can present in public like a debate club or a school assembly.
5. Know that it is okay to be upset or angry and find ways to express your feelings appropriately.

Parents and Teachers

1. Encourage discussion on a variety of topics and listen to their opinions.
2. Teach them to respect themselves. They are important.
3. Don't shut them down or minimize their voice.
4. Help them understand and set boundaries.
5. Help them deal with anger and deal with conflict in appropriate ways.

Moving forward

We have an opportunity to guide our young girls to believe in themselves and state what they want and need. We can help them as they learn to state these wants and needs simply and unapologetically.

Practicing assertiveness at an early age will develop the skill at a time they can think about who they are and what matters to them. This will prepare them to stand up for themselves as they encounter issues through their lives and receive requests for their support for activities, committees and causes that will require their time and attention.

5 BEHAVIOURS OF ASSERTIVE WOMEN

1	*Use assertive body language; hold your head high and maintain eye contact.*
2	*Express your opinions without apologies or caveats.*
3	*Make your requests clear.*
4	*Acknowledge the perspective of others.*
5	*Learn to negotiate.*

Figure 10. From Be-an-Assertive-Woman, wikihow.com, 2023

KNOW HER NAME

Ursula M. Burns
ENGINEER

Ursula M. Burns is renowned for her tenure as the CEO of Xerox, serving from 2009 to 2016, making her the first Black woman to lead a Fortune 500 company. She also holds the distinction of being the first woman to follow another, Anne Mulcahy, as the head of a Fortune 500 company. Burns began her journey with Xerox as a summer intern in 1980 and officially joined the company a year later, following the completion of her master's degree from Columbia University.

U.S. President Barack Obama appointed Burns to help lead the White House National STEM program in 2009, and she remained a leader of the STEM program until 2016.

She has been listed multiple times by Forbes as one of the 100 most powerful women in the world. In 2018 she was featured among "America's Top 50 Women In Tech".[21]

Burns published a memoir, Where You Are Is Not Who You Are: A Memoir, in 2021.

KNOW HER NAME

Jean Jennings Bartik
COMPUTER PROGRAMMER

Jean Jennings Bartik was one of six human computers chosen to work on the new machine called ENIAC–the Electronic Numerical Integrator and Computer. ENIAC is the first programmable electronic, general purpose, digital computer. She and the team taught themselves ENIAC's operation and became its (and arguably the world's) first programmers.

Bartik, who was recognized late in life, became a strong advocate for increased participation by women in Science and Technology.[22]

She was inducted into the Women in Technology International Hall of Fame in 1997. In 2008, she received a Fellow Award from the Computer History Museum. She received the IEEE Computer Society's Computer Pioneer Award in 2009. The IEEE is the world's largest technical professional organization dedicated to advancing technology for the benefit of humanity.[23]

Competitiveness

*Competitiveness—having the desire
and drive to win.*

Girls and boys can demonstrate a drive to win, to be good at something, to succeed and surpass limitations. They have a competitive spirit which drives them to overcome things. They want to be good at something, to be the best, to get the badge, the gold star, the trophy. Being competitive is demonstrated by setting goals and working to be their best.

Goal setting can be incremental—beating your own time or increasing the number of your accomplishments, or larger—shooting for first prize in an event that takes a lot of practise over months or years. Stating the goal, and then setting up a plan to work toward it are great skills to build at all ages.

Teaching our girls to be their best focuses their efforts on things they can control. It helps them to practise, to hold themselves accountable and keep moving forward.

It's also important for us to talk about the outcomes of a competition. Our girls should be proud of their achievements even when they don't win. The ability to create a goal and work toward it is an achievement in itself. It's always easier to not start.

Let's also coach them on how to be gracious as winners and losers. When they win, they should feel proud and comfortable in accepting the win and basking in the positive feeling. When they lose, they should remember and be proud of the effort they put in. They can let their feelings spur them on to try harder or prepare better next time. None of this should take away from the winner's achievement.

Each girl can decide if they want to compete individually or as a team member and whether to compete in academic or physical activities. They can decide if they want to compete within their school or community or at a broader level. Helping them to find the competitions that feel appropriate for them will ensure they don't give up.

Girls will face competition throughout their lives, so it is important for them to learn how to set goals and work to achieve them as we support them to be gracious winners and losers each time.

We have a responsibility as we raise girls to help them learn to compete in ways that are aligned to their personal style. Embrace competition. Feel the joy of it.

DID YOU KNOW?

Since 2020, the **European Innovation Council (EIC)** has introduced several measures to actively promote and support the role of women innovators and researchers to boost Europe's innovation capacity and ensure relevance of its outputs to all citizens. This includes the prioritisation of women CEOs invited to Accelerator Interviews, dedicated initiatives as WomenTech EU to support early stage deep-tech startups funded and led by women, and its Women Leadership Programme to provide coaching and mentoring to EIC-funded women entrepreneurs. The EIC also awards an annual European Prize for Women Innovators, in cooperation with the European Institute of Technology, to recognise and promote women entrepreneurs, who have founded a successful company and brought their innovations to market.

eic.ec.europa.eu
Michiel Scheffer, EIC Board President
Ana Barjasic, EIC Board Member

5 THINGS COMPETITIVE WOMEN DON'T DO

1	*They don't give up when they experience difficulty achieving their goal.*
2	*They don't stop searching for ways to improve.*
3	*They don't see limits as they pass their comfort zone.*
4	*They don't let frustration defeat them.*
5	*They don't forget to have fun while competing.*

Figure 11.

A competitive streak is healthy. We all have one, it just takes different forms depending on our age, our gender, the opinions of others and even what we are competing with, for or against.

Debra   Competition is a good life skill, one that we all must develop to succeed both in school and in our professional careers. We live in a society that values achievement, that celebrates ambition and accomplishment. It is important for girls to understand this and to pursue their ambitions which often involve competition. Certainly there are competitive situtations and even enterprises like sports that are built on competition and require intense focus, commitment and dedication. We celebrate being our best and mastering a skill. The world celebrates anything #1; the best violinist in the world, the Pulitzer prize winner, the Heisman Trophy winner, the Olympian. We use terms like the best of the best, the top of your game, and "elite".

Learning to tap into that part of your spirit is important for girls. It builds their self esteem, enhances their confidence, creates opportunities for them to focus on. Learning helps them find out who they are, what they like and what they might want to do in life. It is also amazing to learn you are good at something. It boosts your spirit.

Competition starts early for both girls and boys. From little league, soccer, spelling bees and gymnastics, to ballet and piano recitals, we are competing at a young age. In fact, as toddlers, we are in constant competition mode—"we can do it ourselves", "we are faster than mommy or daddy running home from the park", "we can get dressed before our siblings". There is a constant aura of competition in our lives, and it is quite positive because it pushes us to achieve something. It creates a streak of independence and competence. We learn to trust our instincts, to push beyond obstacles, to stretch ourselves and perform. Competition is about performance. Competition helps us learn what we are good at and what we want to improve. It could also be the beginning of determining what we want to do professionally, who we become, where we want to focus our energy, what we might want to study. Competition builds our confidence. It strengthens our ability to focus and persevere. It even builds resilience for when we will need to pick ourselves up when we fail, and then try again.

Young girls are fierce, they are determined and can achieve anything they put their minds to. They love to learn; they are excellent builders and collaborators. They take charge and make things happen.

When I was in grade school, I learned about being competitive. I was a decent student, not the top of my class but I worked and studied hard to get good grades. I also got involved in track and field. I was not overly athletic, but I was a good runner, second only to one other girl, Bonnie D. We were in grade seven and she could run like the wind. She was always first and I was second. Sometimes she left me in her dust. I had never seen anyone run that fast. I trained and I tried to increase my speed, but I could never beat her. Then one day our gym teacher/coach decided we were going to run a long race for charity. It was a street race, meaning we were not going to be on the athletic field behind our school. We were going out to run a marathon in the neighbourhood. I had never run a long-distance race before. It was different from the 100-yard dash. When Bonnie D. signed up I did too. We were friends but we were also rivals. Bonnie shot out of the blocks like the wind as she always did. I quickly realized there was no way we could run that race at our usual speed. I quickly got my breathing and my pace under control and hit my stride. I didn't give up. I won the race. My coach was shocked because I was a speed runner, not a distance runner and he had never seen a sprinter win a long-distance race. And from that point on he had me run both. I would never beat Bonnie in a sprint, but she would never compete against me in a long-distance run. She simply could not do it. We remained friends for as long as we went to school together. We were healthy competitors, and there was something special in that. As sprinters we raced hard against each other, and I never stopped trying to beat her. I never did, other than that one long distance run. It was a moment I will never forget. I celebrated it for all that it meant to me, and it encouraged me to try other sports. I knew I was not the star athlete of the school, but I was a strong competitor in the sports I played.

Competition is not about the desire to pummel your opponent or make someone else a loser. It is for you to do your best to compete and win. This could be a spelling bee, a soccer competition or a grade on a test. Competition can be intellectual or physical; it can be an individual pursuit, a team sport or a club activity.

There is a difference between being a competitor focused on doing your best and achieving a win, and the negative side of competition when someone is trying to hurt their competitor.

Do girls and boys start out competing differently? Where we see it, it may be a result of our teaching it. If a girl is not fierce, she may have been coached on what not to do by family or friends to be sure she doesn't make someone else feel bad.

Competition can be spoken of as trying hard and doing your best. There is no negative side to this. There is no intent to cause harm to the other person or team when you are trying to better yourself.

People compete in their careers for a job, a project, a client or an award. It's important for us to communicate to young girls that they might not always win, but they should always do their best. Don't pull back because you are worried about other peoples' feelings. We also need to teach them how to be a humble winner and a gracious loser.

Winning gracefully is celebrating your achievement and not deriding your competitor, and losing gracefully includes congratulating the winner and not declaring the competition wasn't fair or acting hostile to the others.

It's also important to support the girls in our lives to be resilient after a competition that they didn't win so they don't give up and take themselves out of the running for other competitions. Just because they aren't the best yet doesn't mean they should assume they can't be in the future.

We can also instruct girls, and encourage them, to ask for feedback. This can be from coaches, teachers, family, friends and bosses later in life. If we are the coach, we should take time to debrief after giving feedback. Ask the girl what she heard and what they need from us in the future to take action.

The feedback may not always be honest or helpful, but it is still input that can be considered.

I swam competitively through my childhood from age 8 to 18. I placed second behind the same girl many weeks in a row at our meets. I was closing the gap slowly, and yet still coming in second. I didn't lose my drive. I practised hard each day. I focused and did my best each week when we raced. And one day, yes, I beat her! I was thrilled. I was surprised. I was proud. I knew that I earned it through my hard work and dedication. I also knew I needed to be humble and kind. We both jumped out of the pool, and I walked over to her to shake her hand. I knew there would be time later to celebrate at home.

My parents taught me not to take myself too seriously and to have fun while competing. This was not an easy lesson for me as a young girl. I was focused and determined and took every loss to heart. Learning this lesson, however, was essential in order to be prepared to take chances and continue competing throughout my life. They continue in their 80s to compete in golf tournaments and card games and decorating their boat. They enjoy competing even when the winner simply earns temporary bragging rights.

I was recently honoured with a 2023 Woman of the Year award for Women in IT. I was reluctant at first to accept the nomination thinking of all the talented women across Canada who would be in the running for this prestigious award. A small voice in my head reminded me that it was important to believe in myself. The judging took into account demonstrated outstanding business and technology-focused achievements, and also how we have actively sought to bring more diversity to the technology industry.

This was right up my alley. This described my intentions and my achievements. I took the chance and agreed to compete. I was surprised when I won, and it took me a few days to share the news. And then I remembered what I tell other people. Celebrate your achievements.

We need to help girls to own their wins. Girls can have a tendency to apologize or diminish their win by saying they only won because the other person made a mistake. After a boy wins and someone congratulates them, they might say "Yes, I killed it!" and a girl might say "Oh, I got lucky."

Let's remind girls to remember they prepared, gave it their all and won. Let's teach them to do the celebration dance when they win.

Let's hear from the girls

Meet Shiku, age 9

She is interested.

Shiku became interested in technology at the age of two watching kids on YouTube. She got her first iPad and watched educational videos and learned to compete as she played games including Minecraft and Subway Surfers.

Shiku likes playing softball, anime and is interested in fashion design.

Her mom is in Finance and Accounting for an investment firm. Her dad is in Finance and Technology for financial services.

She feels the power of tech.

She uses the YouTube channel for ideas and has started to work with the Pro-Create software. She is creative and interested in

Digital Art and Fashion. She sews and uses 3-D printing to build samples for her sewing.

There are no computer programs at her school and no robotics club, so she is finding her way through her own curiosity and interest. She thinks about how we can get more girls interested in tech. We need to think about the girls who don't want to play games or code, but they want to use technology for creative purposes.

She is taking action.

Shiku enjoys interacting on social media. She quickly comments that girls need to create their own platforms.

When asked what she wants to do with her knowledge, learnings and experience with technology, Shiku replies that she would find ways to reduce hunger and clothe the homeless.

let's hear from the girls

Meet Riya, age 14

She is interested.

Riya became interested in technology at the age of 4. She was exposed to tech as she started taking pictures and playing games on her mom's phone. She now uses a Surface Pro and an iPhone. She exhibits her competitiveness with her examples and interest in robotics competitions.

Riya started using Scratch, a programming language, in grade two. In grade five, she was introduced to Robotics at school and thought the robots were cool. She enjoyed learning to make them move in different directions and participating in competitions. She learned to use Lego Block language, and then moved onto Python and MICROP which is a text-based language where she learned about using functions and "if/then statements".

She loves to play basketball and plays the trumpet. She worked with her mom during Covid on a special project. They created an Instagram account and posted a Math problem or question every few days, and then solved it with a videogram.

Her mom is a Senior Scientist—Biostatistician, a Cancer Research Program Lead and a full Professor. Her dad is a Computer Engineer working on machine learning.

She feels the power of tech.

She participated on a team in the First Lego League (FLL) innovation project. The theme was around energy. Her team used a stationary bike and demonstrated how you could exercise on it to create energy that was then stored in a computer battery. The competition measured the teams on innovation, energy creation and personal values including how well they worked together as a team to solve problems. There was blind voting used to make the outcomes fair. She participated in regional championships and the West Provincial championships.

In FLL competition, community groups can participate so it is not always a school team. There are no barriers to entry, and there is a maximum of 10 people on each team. Her team has an equal mix of girls and boys.

There are three levels to FLL to encourage building on the previous learnings:[24]

- Discover (ages 4-6), they are introduced to the fundamentals of STEM.
- Explore (ages 6-10), they use their background knowledge of STEM and put it into practice as they work in teams to design and build robots.
- Challenge (ages 9-16), they apply their STEM skills combined with critical thinking to work with a team, build a robot and compete.

She is taking action.

Riya is participating in activities to learn more about technology itself and the activities that leverage technology. She thinks we could get more girls in STEM by creating all-girl robotics clubs. These would appeal to girls who might otherwise be hesitant to join. We need to continue to challenge the stereotype that only boys are interested in technology so girls don't miss opportunities for the wrong reasons. We can highlight hobbies girls like and highlight examples of the technology that can be used related to these hobbies.

When asked what she wants to do with her knowledge, learnings and experience with technology, Riya confidently states we have great potential to leverage technology to diagnose medical issues faster.

Tips and Techniques

Girls

1. Discover / embrace what you like to do.
2. Try new things and challenge yourself to build confidence.
3. Do not let failure discourage you from competing.
4. Find your support system—friends, family, schoolmates and coaches.
5. Enroll in school or extracurricular competitions—robotics, team sports, individual sports and even writing contests.

Parents and Teachers

1. Encourage them to find out where they excel.
2. Expose them to many interests: intellectual, physical and creative.
3. Give them opportunities to develop their skills.
4. Talk about competition—winning, losing, failing and getting back up. Celebrate the wins, see failure as fuel.
5. Ensure you are encouraging an interest of theirs, not yours.

Moving forward

We have an opportunity to encourage our young girls to establish goals and work hard to be their best. We can ensure they understand it is okay to have the drive to win, to surpass limitations and expectations, and to win.

Competing at a young age will develop the skill at a time they can enter different competitions and learn what they are good at, as well as where they enjoy spending time to increase their skills. This will prepare them to try new things, to understand they need to put in effort to get better and to celebrate the wins achieved.

8 WAYS TO BE MORE COMPETITIVE

1	Know the game never ends.
2	Always give 110%.
3	Compare yourself only to yourself.
4	Get in touch with your dark side.
5	Differentiate yourself.
6	Learn how to lose.
7	Never make excuses.
8	Give yourself credit when credit is due.

Figure 12. From 8 secrets to being way more competitive, edlatimore.com, 2024

KNOW HER NAME

Annie Easley
COMPUTER SCIENTIST, MATHEMATICIAN AND ROCKET SCIENTIST AT NASA

Annie Easley worked for the Lewis Research Center (now the Glenn Research Center). She also worked with NASA and the National Advisory Committee for Aeronautics. Her 34-year career included developing and implementing computer code that analysed alternative power technologies, supported the Centaur high-energy upper rocket stage, determined solar, wind and energy projects, identified energy conversion systems and alternative systems to solve energy problems. Her energy assignments included studies to determine the life use of storage batteries, such as those used in electric utility vehicles. Easley was a leading member of the team that developed software for the Centaur rocket stage. This work helped lay the technological foundations for future space shuttle launches and launches of communication, military, and weather satellites.[25]

KNOW HER NAME

Mary Wilkes
SOFTWARE DEVELOPER

Mary Wilkes is noted in the field of Computer Science for the design of the interactive operating system LAP6 for the LINC, one of the earliest systems for a personal computer. She is also known for being the first person to use a personal computer in the home. Her work has been recognized in Great Britain's National Museum of Computing's 2013 exhibition "Heroines of Computing" at Bletchley Park, and by the Heinz Nixdorf Museums Forum in Paderborn, Germany.

Wilkes left the computer field in 1972 to attend the Harvard Law School. She practiced as a trial lawyer for many years. In 2001, she became an arbitrator for the American Arbitration Association, sitting primarily on cases involving computer science and information technology.[26]

Perseverance

Perseverance—persistence in doing something despite difficulty or delay in achieving success.

Girls and boys have the power to persist in their efforts when success is not immediately achieved. We can look for these moments, congratulate them when we notice and cheer them on as they persevere. Perseverance is important in tech when trying to achieve a longer-term goal that may seem out of reach due to the complexity or worry about being able to complete the work required. This includes stepping into the unknown, dealing with disappointment and treating setbacks as a distraction.

Stepping into the unknown is critical in tech as you explore new ways to accomplish things and new ways to think about problems. This will allow you to solve existing problems that seemed insurmountable and to innovate with new features and functions.

Dealing with disappointment is important in technology as you learn from the trial-and-error process. The first thought or important first step in creating is to get started. The ability to keep going, to keep thinking, to overcome the challenges as they occur is what allows you to experiment and innovate, to improve incrementally and radically on what already exists and on your own ideas.

Setbacks can be painful. They can also be empowering if you think of them as a temporary distraction. A setback provides the space to learn and to gain a deeper understanding. You can assess what parts of your expectations and beliefs were true, and where they took a turn that proved incorrect. You gain evidence and perspective from this setback that is useful as you move forward.

We have a responsibility to girls to help them develop these skills through everyday experiences so they build the perseverance that will be valuable throughout their lives.

DID YOU KNOW?

Homeward Bound is a ground-breaking, global leadership initiative set against the backdrop of Antarctica, which aims to increase the influence and impact of 10,000 women leading with a STEMM (Science, Technology, Engineering, Mathematics, and Medicine) background in making decisions that shape the future of our planet.

Homeward Bound similarly "provides women with leadership, visibility and strategic skills, a sound understanding of the science, and a strong purposefully-developed network enhancing their ability to lead for the greater good, to impact decision-making for a sustainable future".

homewardboundprojects.com.au

Founders: Fabian Dattner, Dr. Justine Shaw, Professor Mary-Anne Lea, Dr. Jess Melbourne-Thomas

5 THINGS PERSEVERANT WOMEN DON'T DO

1	They don't lose sight of their goals.
2	They don't let worry stop their progress.
3	They don't reinvent the wheel when there is opportunity to reuse assets.
4	They don't expect to fail.
5	They don't miss the opportunity to prioritise impactful tasks by focusing on minor tasks.

Figure 13

Girls and boys can demonstrate intense focus. Just watch young children playing, doing a puzzle, drawing, creating. They are laser-focused on the outcome and what they are trying to achieve. Try interrupting a toddler focused on doing something. They barely hear you; they might even ignore you. They are focused. They are trying to complete or create something that is important to them. Watch them colour, or build something or watch television. Intense focus.

Achievement might be the underpinning of perseverance and focus. It is the result of our innate desire to overcome, to do something we are interested in or didn't think we could do. Are we responding to a challenge or encouragement?

We persevere from the time we are young without being conscious of it, possibly because we are trained to not leave things unfinished. We hear early in our youth to not start something you cannot complete. When you sign up for activities your parents tell you to see it through, even if you don't want to. We use words like "don't give up, you cannot quit, but you must see it through, your classmates or your team is depending on you". It is always there, underneath the surface in everything we do. Keep doing, keep pushing, keep striving, keep moving things forward.

We rarely say stop, or you can give up, or you can change your mind in midstream. You can change your mind going forward and maybe not choose that activity the next season or following year, but we expect completion. So interesting that we don't like to fail to see something through. Achievement is a core value that many girls live by, and it is applauded in our society. We celebrate great things. We cheer people on when they see something through. We encourage our children to stretch themselves, to reach higher, try harder and develop their skills, whether it be athletics, music, reading or creating. We persevere because we are always trying to raise the bar.

Achievement is one of the top values I live by. This was reinforced by a "values" exercise I did last year. I was raised to do many things. I was encouraged to participate in everything I could: Girl Guides, dance lessons, narrating the Christmas pageant at church, joining the debate club in high school. It was about trying new things, expanding my interests and learning as much as I could. Not everything was easy, but I continued on. I didn't give up just because I might be struggling a bit. Sewing taught me that. I learned to sew at 10 years old. My grandmother used to sew, and she taught me a few things. I would make napkins or something simple. She encouraged me to keep trying new things to sew, and she even made a deal with me. If I wanted to make my own clothes, she would buy the fabric.

I was excited about the possibilities in our Home Economics class when we had to make an apron. It was supposed to be simple, but I decided to make a double sided/reversible apron of pink and green plaid fabric, with tabs at the side with buttons and buttonholes.

When I went into class with my fabric and pattern, my HomeEc teacher chastised me for picking something a novice could not make. She asked me why I had made that choice. My response was "If I am going to be taught to do something, and have help from someone who does sew, why would I pick something easy?" She smiled and said, "All right then, I will teach you." And it was not easy. I was never so proud of anything and a lifelong love affair with sewing began. I am still sewing over 55 years later. I made everything from my wedding dress and all the dresses for my bridal party and continue to enjoy sewing clothes for myself, my children and grandchildren and also curtains, bedding and other home décor. This creative outlet brings me joy.

please **JOIN** THERE'S NO PLACE LIKE TECH

Kelley

Learning a new skill and being proud of the achievement often comes from incremental progress and perseverance. The ability to continue to be persistent is difficult and can be enhanced when you are able to see the goal in front of you.

Frustration is typically felt while being persistent. Knowing the goal, seeing it and feeling it, helps you remember and stay connected to why you are continuing to try.

I remember learning to ride a bike. I was practicing on a hill, falling off, and even getting hurt and cut and bleeding. None of this deterred me because I could feel the goal in my soul. I wanted to ride because my brother rode his bike, and I wanted that same sense of freedom that he was experiencing, under my own power, following my own target, and at my own pace.

All this was reliant on other people allowing me to be frustrated and continue to try and figure things out myself. These little wins as a young person helped me build the muscle for perseverance, and then take on medium goals as I grew, and then larger goals. I remember that day in detail and am thankful for learning I had the ability to overcome the frustration. I built trust in myself, and

I can still feel the exhilaration that occurred once I demonstrated the new skill of riding my bike.

Overcome obstacles and achieve success, then polish the skill over time. Be open minded and willing to listen to feedback and adjust based on the perspective of others who have had experiences in their lives.

The ability to try again is pronounced in young children as they attempt to learn things that someone else may be doing, or something they can envision in their heads. We need to encourage the girls in our lives to continue this as they get older, so they don't question their ability to persist based on the opinions of others.

Tracking progress is helpful to see how far you have come. We can mention this to young people by reminding them of the incremental steps they have taken and the growth we can see. We can also teach them to track these changes on paper or in an app. Seeing their achievements accomplished over time also reminds them of the perseverance they have already demonstrated.

For skills you are trying to learn, there is also the need to accept the curveballs thrown at you in the process. These might be your own physical or intellectual constraints, additional steps that you didn't understand, or external factors that change during your learning.

When Debra and I decided to write our first book, we thought we had done enough research to know what we were getting into. We came to understand over time that we had only scratched the surface of learning about the process to author a book. We had to learn about editing, publishing and the launching of a book, and the activities involved in those processes that were completely unfamiliar to us. We also encountered the global pandemic in the middle of the process. This impacted our plans to meet in person and work together, to interview women in tech at events we typically attended, and to launch the book with a party to thank all the people involved. We persevered and laughed along the way, recognizing our ability to learn and pivot and accept help from many people with experience that we did not have. It was a fantastic journey that didn't look much like we had thought it would but was more rewarding than we could have even hoped.

It's important for us to teach girls that frustration is normal when you are learning new skills, and to continue trying. It's also important for us to model this behaviour so they may understand what it looks like.

let's hear from the girls

Meet Charlotte, age 13

She is interested.

Charlotte became interested in technology at the age of three. One of her early experiences was playing games including Minecraft. She thinks parents need to get kids involved in tech early by exposing them as young as possible. She thinks sometimes girls don't get involved with tech because it is an unfamiliar territory. This can occur if boys get more exposure at an early age to tech than girls do. She understands that persevering allows you to progress in your skill.

Charlotte likes Lego and competitions. She describes herself as positive and curious.

Her mom is a Physician. Her dad is an Accountant.

She feels the power of tech.

She got involved in both typing and coding classes. She commented that girls are playing many games, with Fortnite and Roblox being two favourites. She thinks about cyber bullying and online safety and how kids need to be educated on these risks.

She has started making her own games and 3D designs using basic programming. In grade three, she was exposed to a variety of robotics assembly kits.

Her school has multiple assemblies a year and the winter assembly focuses on robotics providing awareness of the program to all kids at the school. She attended a summer camp in grade five that exposed her to robotics, and that was how she got involved in a robotics club. She is now on a team that is competing in the First Lego League (FLL) competition.

The robotics team builds its own robots. It is a commitment of time to your team members. There are four parts to the competition that are all equally important—building the robot, the robotic mission, an innovative design, and teamwork.

She is taking action.

Charlotte and her robotics team presented to the grade 3-5 students at their school in the 2023 school assembly. The presentation was about the innovation project (Z-Pipe) they had completed the previous year. Their goal was to inspire students to join FLL and become interested in STEM, just as it had done for her in grade three.

When asked what she wants to do with her knowledge, learnings and experience with technology, Charlotte says she is confident we can achieve net zero emissions with the right attention and perseverance. She also thinks of a day we will leverage innovative ideas for packaging and no longer require materials like bubble wrap.

Let's hear from the girls

Meet Gaby, age 8

She is interested.

Gaby became interested in technology at the age of four. She started by playing games on her mom's phone. At age seven, she discovered she could use technology to create games herself. She used Kodable, a platform that is designed to help kids learn coding through interactive games and self-guided activities. She learned to build pathways.

Gaby likes to play Math games and hopes to use technology as part of her job in the future. She likes Science and Social Studies and enjoys learning about other places in the world. She is learning French and Chinese. She is also a competitive Rhythmic gymnast and has earned several medals. She demonstrates perseverance as she increases her skill level in each activity she takes on.

She wants people to know she is really funny, and she loves her brothers, especially Archibald who is 12 and closest to her age. She likes pizza parties and secret Santa.

Her mom is a Marketing professional, a Chef and a Christian recording artist. Her dad is a Chief Information Officer.

She feels the power of tech.

She quickly pointed out they don't use computers at her school, they use iPads. There are girls and boys involved in the Kodable online learning at her school, with more girls than boys involved. She uses this online platform to practise her coding.

She believes technology tools and applications should be for kids of any age. She believes anyone can use technology and that you need to keep trying when something doesn't work. She has learned to use the Scratch programming language. This is a high-level block-based visual programming language and website aimed primarily at children for use as an educational tool.

She also enjoys playing games on her Nintendo Switch. Super Mario is her favourite.

She is taking action.

Gaby has added to her tech skills learning Code Ninja, Scratch and JavaScript. Using these skills and her perseverance, she built her own game where you can earn belts and receive stars based on your actions.

She thinks we need to create more games for girls that they will like and says not all girls like fighting games. She likes Fortnite and good sporting games like Mario Olympics. She also enjoys the Oubey Mindkiss project.[27] This is an interactive, intercultural and experimental global platform aimed at making the work of the artist, Oubey, posthumously accessible to a broad public.

When asked what she wants to do with her knowledge, learnings and experience with technology, Gaby confidently stated she wants to be a surgeon specializing in surgery for kids, a gymnastics teacher and work for the government.

Tips and Techniques

Girls

1. Commit to your goals.
2. Do not be discouraged.
3. Be prepared with a plan for when you get stuck.
4. Ask for help.
5. Do not seek perfection.

Parents and Teachers

1. Ask questions to understand what they are trying to do.
2. Seek additional resources for help. You don't need to be an expert.
3. Don't let them quit.
4. Encourage them to take a break when frustration sets in.
5. Help them set reasonable goals.

Moving forward

We have an opportunity to recognize the perseverance in girls when they persist, learn, try again and eventually succeed. Every time they learn something new, we can point it out and celebrate their efforts along with their accomplishments.

Learning to persist until they achieve their goals builds the skill of perseverance that will benefit them in attaining personal and professional goals at all ages.

7 STEPS TO TEACH KIDS PERSEVERANCE

1	Take your child out of their comfort zone.
2	Let your kid get frustrated.
3	Model a growth mindset.
4	Brainstorm together.
5	Teach that failing is okay.
6	Praise effort, not accomplishments.
7	Be a gritty parent.

Figure 14. From 7 Steps Parents Can Take to Teach Kids Grit, verywellfamily.com, 2024

KNOW HER NAME

Hedy Lamarr
THE MOTHER OF WI-FI

Hedy Lamarr may initially be remembered as a beautiful Hollywood actress. The screen star was born in Vienna as Hedwig Eva Maria Kiesler. But there's much more to Lamarr than meets the eye.

During World War II, Lamarr read that a radio-controlled torpedo had been proposed. However, an enemy might be able to jam this torpedo's guidance system and set it off course.

Lamarr, along with George Antheil, developed a radio-based torpedo guidance system that was immune to jamming using a frequency-hopping signal. Initially few people took the actress seriously, and her patent eventually expired without being used in the real world. This technology went on to be used in many of our essential technologies today, including Wi-Fi and GPS.

In 2014, Lamarr and Antheil were inducted into the National Inventor's Hall of Fame.[28]

KNOW HER NAME

Karen Sparck-Jones
COMPUTER PROGRAMMER

Karen Sparck-Jones was a self-taught computer programmer during a time when there were *very* few female programmers in the field. Her most notable contributions centred around inverse document frequency and index-term weighting — two big concepts that helped create the modern search engines we have today.[29]

In the 1980s, Spärck-Jones began her work on early speech recognition systems.

She received numerous awards including the Gerard Salton Award (1988), Association for Information Science and Technology (ASIS&T) Award of Merit (2002), and Association for Computational Linguistics (ACL) Lifetime Achievement Award (2004).[30]

Confidence

Confidence—a feeling of self-assurance arising from appreciation of your own abilities or qualities.

Girls and boys can build confidence as they learn to trust in themselves and in their own abilities and knowledge. They can challenge themselves and celebrate wins, large and small.

Building trust in yourself and in your own abilities is similar to learning to trust others. Start with understanding yourself and acknowledging where you have doubts. Change the internal dialogue to create a compassionate friend—yourself. Instead of thinking about what you can't do, think about what you can do, what you have learned and how hard you are able to push yourself to get better when you are motivated to learn new skills.

Remind yourself it's okay to feel vulnerable. It is brave to acknowledge uncertainty or imperfections in your skills and to take action to get better and demonstrate the progress to yourself.

When you achieve wins, large or small, take a moment to think about the achievement. Smile to yourself, share the win with a friend, write it down in a journal or in your calendar and say to yourself, "I did that!"

We often spend our thoughts, time and effort on the things we aren't yet able to do. Make sure you also spend time on the things that are going well, the things you have achieved and the moments of triumph.

Appreciate yourself. You are worth it.

We have a responsibility to girls to help them identify what matters to them and what they want to achieve. This will help them to be confident as they show their authentic self and all their accomplishments to the world.

DID YOU KNOW?

Women in Tech Africa (WITA) was founded in 2013 to Impact their Communities Positively. WITA believes that women are equally capable of being at the forefront of technological development and advancement in Africa and the world at large.

Women in Tech Africa's Vision:
- Creating today's female leaders and role Models for tomorrow's Women
- Showing the world what a strong African woman is capable of achieving
- Support African growth through technology

Women in Tech Africa is the largest group on the continent with membership across 30 countries globally with 12 Physical chapter in Ghana, Nigeria, Malawi, Zimbabwe, Somalia, Germany, Ireland, Britain, Kenya, Tanzania, Mauritius, and Cape Verde. Women in Tech Africa is also the 2018 recipient of the United Nations Equals Award for Leadership in the Women and Technology Space.

womenintechafrica.com

Ethel D Cofie, CEO and Founder

8 THINGS CONFIDENT WOMEN DON'T DO

1	They don't doubt themselves.
2	They never listen blindly.
3	They don't cave to peer pressure.
4	They don't ignore their instincts.
5	They don't glorify busy.
6	They don't find silence uncomfortable.
7	They don't want fans, they want supporters.
8	They don't equate who they are with what they have.

Figure 15. From 22 Things Confident Women Don't Do, lifehack.org, 2016

Confidence is a life-changing skill for girls. Girls are under pressure in many aspects of their lives. There are many factors that can impact their confidence. The social demands on girls are tremendous. They are being compared to unrealistic images, sizes, expectations on how to behave and what to be. This pressure can unravel even the most confident of women. Building their confidence is important. Girls start their lives confident, bold and brash and don't let anything stand in their way. They are unstoppable. Then they let the voices of others start to sneak in, and that can erode their confidence. They let opinions of others sway them. They question their own judgement because someone challenges them. They are concerned that people won't like them. They feel judged and criticized if they do what they actually want to do. Their confidence takes a hit. We need to prevent that from happening and block those influences that may attack their own sense of self.

As toddlers start to expand their circle at school and in community events, their areas of interest expand, and their confidence grows. It is at a certain point in their adolescence and teenage years that we see their confidence start to waiver. The influence of others, the school environment, social media and many other factors are potential confidence busters. It will take effort to put those influences in their proper place to keep them from whittling away at the strong sense of self that girls have. It is important for them to hold on to every ounce of their confidence. It will serve them well throughout their entire life.

Girls are amazing. Girls are funny and smart. Girls are confident. Let's keep it that way.

When I was 15-16 years old, I had a boyfriend, a downtown boy, Jesse. I met him at a dance at my church. I was not really allowed to have a boyfriend. Jesse was handsome and a bit of a bad boy, mature well beyond his years. He was raised on the streets of Montreal with a very different upbringing than my own. As I was book wise, he was street wise and the difference between us was palpable. Jesse had a former girlfriend, one of the most beautiful girls I had ever seen in my life. She looked like the singer, Andra Day, with those amazing green eyes. At one of the dances at the church I came out of the washroom stall, and there she was with her posse, about five other downtown girls. That meant they were tough, and a little rough. At that moment I knew I was in big trouble! They were there to hurt me, and by that I mean beat me up. She was angry that Jesse was now with me, and I was scared because this bathroom was very small. I could imagine her slamming my face into a piece of porcelain, so I had to think quickly about what I was going to do. They were blocking the door and no way were they going to let me just walk away.

I slowly went to the sink, washed my hands, and stood there as she threatened me. I was calm; I was self-assured. I asked her to come to the mirror and look in it. As she did, I said to her, "Do you see how beautiful you are? You're the most beautiful girl in this church hall and you want to fight me over a boy?" I paused and let that sink in. Then I said to her, "If you want to hurt me then do what you have to do, but if you hit me, you better make sure I stay down. But before you do, if you want Jesse so badly you can have him. I will never fight over a boy. And you shouldn't either. You are more than that and can have any boy in this place. No boy is worth demeaning yourself for, but if you want him so badly, go get him." And much to my surprise and relief she stepped aside and let me leave. I knew at that moment who I was, strong, confident and brave. You can be too.

True confidence is a feeling, not an action. We often think of people as confident when we see their behaviour. This can be loud talking or a swagger in their walk. We need to remember that people acting in a way to project confidence does not indicate they really are confident. Confidence is that feeling of self-assurance, knowing you can do something and appreciating your abilities.

I feel confident my friends and family have my back. This is an area of confidence that allows me to take risks knowing if things go astray, there is someone there to catch me. This is something we can teach young girls. Pick friends who will be there for you in good times and bad. These aren't people who just listen, but people who will provide advice at the right time and a shoulder to lean on at other times. Having a safe place to turn enables us to build our confidence with a spirit squad to cheer us on.

I was confident as a student. I listened, studied, and I cared to learn. I knew that I could do well in school because I put in the work, and I asked for help when needed. This didn't mean I was great at every subject. I excelled at Math and Science and was challenged by Foreign Languages and Art class. I realized I was creative, but not necessarily artistic. Knowing this about myself allowed me to confidently make choices about my future. I focused career opportunities on my greatest strengths without giving up completely in the areas where I was less successful. I continued to try my hand at art and have provided humorous moments for my family with the objects I created.

Debra and I were encouraged by many people to make videos about the topic of women in technology. We had this input and encouragement for three years, and we agreed that it would be a good goal for us, but we were resisting without even talking about why we were hesitant. I was personally uncomfortable because I hadn't done it before and didn't even know where to start. The best advice we got was just to start, and if we weren't okay with it then we didn't need to share the videos. We had opinions and experience to share, so time and content weren't the issue. It really was about being comfortable and feeling confident to start something new that neither of us had experience doing. We realized we were overthinking it. We now meet every week, and video ourselves talking about these topics.

That feeling of self-assurance needs to be practised in social situations. From the time we are young, there are moments when we can doubt ourselves and our abilities. I remember that funny feeling I got when meeting new people, when starting middle school, moving to a new city in grade nine, starting my first job in high school and moving away to college. Each of these moments created opportunities to meet new people. I was excited and cautious. I didn't know them. They didn't know me. Will we "click"? The truth is I realized not everyone would become a close friend, but some people would. Being confident in our abilities to excel at a task is different from being confident in our abilities to connect with people. Both aspects of confidence are important and rewarding if we just take the chance.

My mom calls me fearless. My dad calls me courageous. This all stems from confidence. I know my ability to be confident and appear fearless and courageous is built on a solid foundation of pushing myself to learn and a family who celebrates the effort and the achievements.

We want girls to exhibit and acknowledge the strong skills they have, and to keep exhibiting and build on them. It's important for us to demonstrate that by pushing ourselves to continue building our skills and appreciate our own abilities and qualities.

As we develop these skills, it's also important to be proud and own the successes. We don't want our girls to focus on what they wish were different and what they don't like about themselves. We want them to celebrate the abilities and attributes they have and the progress they make.

Celebrating your abilities is not being conceited. It's just confidence.

let's hear from the girls

Meet Zoe, age 6

She is interested.

Zoe became interested in technology capabilities at age two. She took her mom's phone and started taking pictures. She likes to post photos and videos on YouTube, download videos and film her mom while she is cooking. She also wants us to know she likes milkshakes.

Zoe displays her confidence as she mentions that she finds Math easy and is ahead of everyone in her class. After saying this, she quickly adds that she might regret that answer. If her teacher knew how easy it was for her, she might get harder work.

Her mom works in Fintech. Her dad works in Technology.

She feels the power of tech.

She talks about technology in a way that you understand it as an integral part of her life. She plays simulation games like Sims and enjoys finding interesting Apps that are free. She is interested in games like Toca Life because you get to be the boss in this game. So far, she is not planning to pay for her games.

She had an Amazon tablet that was geared toward young children and made use of it for years. She now feels, at age six, she has outgrown the functions available on this tablet. She is currently using a Samsung tablet that has more functionality.

She uses Sketchbook, a web program, which allows you to colour whatever you like online.

She plays soccer, and her parents stress that she needs to include both technical and physical activities in her week. She is currently thinking of taking karate and says with a smile that she is ready to take on anyone.

She is taking action.

Zoe talks to her parents about what she wants to do. Her dad is going to build a custom desktop computer for her that meets her growing needs and desires.

One day she'd like to own a shoe company.

When asked what she wants to do with her knowledge, learnings and experience with technology, Zoe says she wants to put technology to good use. She would combine various technologies together to create something new and exciting. She tells us what she will create, but says we'll have to wait and see it in the future.

Let's hear from the girls

Meet Adrianna, age 11

She is interested.

Adrianna became interested in technology capabilities at the age of eight. She was interested in STEM activities where you could make a circuit. She used KiwiCo that allows you to generate code quickly.

Adrianna learned Scratch, a coding language with a simple visual interface that allows you to create your own stories and games. She has used this to post blog entries. She enjoys competitive gymnastics and acro dance, a style that combines classical dance technique with precision acrobatic elements. She displays her confidence as she easily speaks of technical capabilities and her creative endeavours.

Her mom is in Energy Transition. Her dad is in Technology supporting Financial Services (fintech).

She feels the power of tech.

She sometimes calls herself "Dad's assistant". He has exposed her to coding on tablets and phones. She recently helped with the Dads4Daughters STEM website using the WIX platform and tools to publish. WIX is a code-free website builder.

She joined a Robotics club at the library she visits before school. She attends Spirit of Math, an after-school Math education program and talks about using ChatGPT to learn approaches for solving difficult Math questions. She used ChatGPT to help her family plan their vacation to make the most of each day and effectively schedule all the activities they wanted to experience.

She is taking action.

Adrianna realizes we don't need to learn just by reading about a topic. She has experienced learning through participating in a hands-on activity and this helps her understand how it really works. By doing so girls can relate to tech in a way that is connected to their interests.

When asked what she wants to do with her knowledge, learnings and experience with technology, Adrianna thinks about the real problems caused due to the changes in weather. She believes we can use technology to address some of the climate change issues we are experiencing in the world.

Tips and Techniques

Girls

1. Know what you are good at and embrace it.
2. Be proud of yourself and each accomplishment.
3. Don't compare yourself to others.
4. Don't doubt yourself. Listen to the voice inside you.
5. Your self-worth should not be dictated by someone else's opinion.

Parents and Teachers

1. Listen to what they are interested in and embrace it.
2. Highlight what they are good at and mention it often.
3. Pay attention to the words they use to describe themselves and challenge the negative ones.
4. Don't underestimate the circles they travel in.
5. Expose them to strong, confident women including your family, friends, and community.

Moving forward

We have an opportunity to encourage the confidence in girls when they display their talents and abilities in private and public settings. We can counter any negative self-talk we hear and, equally important, we can amplify the moments of self-assurance they experience.

Learning to believe in themselves is critical at every stage of their life to participate fully, to ask for what they want and need and to enjoy their personal efforts, talents and accomplishments.

7 SIGNS OF A CONFIDENT WOMAN

1	Independence and Self-Reliance.
2	Positive Body Image.
3	Assertiveness.
4	Emotional Maturity.
5	Self-Worth.
6	Knows Her Strengths and Weaknesses.
7	Adaptability and Resilience.

Figure 16. From 7 Signs of a Confident Woman You Need to Know by Christel Owoo

KNOW HER NAME

Dr. Marian Rogers Croak
ENGINEER

Dr. Marian Rogers Croak collaborated with colleagues while employed by Bell AT&T Labs to create VoIP (Voice over Internet Protocol) Technology. VoIP is an internet technology that permits telephone use over the internet. Verbal and visual communications on this platform leverage internet transmission instead of conventional telephone cable connections. Her work has furthered the capabilities of audio and video conferencing. VoIP technology is vital for remote work and conferencing today.

Dr. Croak is now Vice President of Engineering at Google. She holds more than 200 patents. She was inducted into the Women in Technology International Hall of Fame in 2013. In 2022, Croak was inducted into the National Inventors Hall of Fame for her patent regarding VoIP.[31]

KNOW HER NAME

Gladys Mae West
MATHEMATICIAN

Dr. Gladys West was inducted into the Air Force Space and Missile Pioneers Hall of Fame during a ceremony at the Pentagon. Lt. Gen. David Thompson, Air Force Space Command vice commander, presented Dr. West with the award for her decades of contributions to the Air Force's space program. This is one of Air Force's Space Commands Highest Honors. Hired in 1956 as a mathematician at the U.S. Naval Weapons Laboratory, West participated in an astronomical study that proved, during the early 1960s, the regularity of Pluto's motion relative to Neptune. From the mid-1970s through the 1980s, she programmed an IBM 7030 "Stretch" computer to deliver calculations for a geodetic Earth model. The algorithms West used accounted for the variations in gravitational and tidal forces that distort Earth's shape. The geoid optimized what ultimately became the Global Positioning System (GPS).[32]

Final Thoughts

Girls Belong in Tech.

Girls are likely to show their interest in technology at an early age. We may need to encourage them if it does not yet feel interesting to them. It is important for us to find diverse ways to expose them to technology and provide learning experiences that can help them make informed decisions. We also need to ensure they don't opt out of classes early in life that create unintended consequences that hinder their future options.

Whether girls want to be "in tech" or not, their lives will be influenced by tech, and they will need to understand how to use it. Creating tech—or just being an educated user and beneficiary of the available technology—are both reasons we need to expose them to the world of tech. We can challenge them so they advance from being a casual user of technology to potentially becoming an inventor of technology or an innovator leveraging technology.

There are many career paths in tech to be explored. When we look at women who have succeeded in tech careers, their backgrounds include diverse studies. The tech industry requires people with courage and confidence in addition to education and experience.

We are inspired by the stories from girls who show an interest and an aptitude in technology at an early age and women who have changed our world with technology innovations. These innovations have formed the base for many products and services we use in our daily lives.

These innovators persevered and more than that, they paved the way for others. They established paths for girls to have better access to courses and information as they learn to code or design new products. They didn't take no for an answer.

They stood tall. They stood strong. They believed in themselves, and they believed in a future generation of girls stepping forward to make great contributions.

This is the time girls. It's yours for the taking. Let's make it happen.

Encourage girls to consider tech careers.

Support them as they face challenges.

Remind them of their many talents.

Celebrate them as they succeed.

We need them to join!

Resources

STRENGTH

1. discoverymood.com/blog/ characteristics-of-mentally-strong-women/
2. righttoplay.ca/en-ca/national-offices/national-office-canada/stories/the-incredible-strength-of-a-girl/
3. girlscouts.org/en/raising-girls/leadership/life-skills/how-can-girls-be-strong.html
4. forbes.com/sites/amymorin/2020/02/11/7-things-mentally-strong-women-believe/?sh=a93c0d41b4e1
5. marymount.sudburycatholicschools.ca/regals-find-strength-in-the-power-of-being-a-girl/

COURAGE

1. linkedin.com/pulse/importance-courage-development-girls-candace-doby/
2. candacedoby.com/cool-girls-guide-to-courage/
3. outsideonline.com/culture/active-families/10-ways-raise-brave-girls/
4. linkedin.com/pulse/10-habits-courageous-women-sonia-mcdonald/

5. giantleapconsulting.com/courageous-leadership/six-habits-of-courageous-women/

INDEPENDENCE

1. sciencedaily.com/releases/2015/08/150817085619.htm
2. thestartupsquad.com/6-ways-to-teach-your-daughter-independence/
3. basicsbybecca.com/blog/independent-women
4. lifehack.org/articles/communication/12-things-strong-independent-girls-dont.html
5. parent.com/blogs/conversations/11-ways-to-foster-independence-in-your-kids

LEADERSHIP

1. girlsleadership.org/
2. guides.womenwin.org/ig/programme-design/developing-girls-leadership
3. girlsthatcreate.com/leadership-in-young-girls/
4. plan-international.org/publications/taking-the-lead/
5. canadianwomen.org/the-facts/women-and-leadership-in-canada/

ASSERTIVENESS

1. cultureplusconsulting.com/2018/03/10/gender-bias-work-assertiveness-double-bind/
2. bethanywebster.com/blog/female-assertiveness/
3. goodtherapy.org/blog/Strong-Like-Amanda-Teaching-Girls-Power-Assertiveness
4. rootsofaction.com/assertive-confident-girls/
5. kidshealth.org/en/teens/assertive.html

COMPETITIVENESS

1. psychcentral.com/relationships/competition-among-women
2. statepress.com/article/2021/02/\specho-girls-competition#
3. hbr.org/2019/11/research-how-men-and-women-view-competition-differently
4. irishtimes.com/culture/books/why-it-s-wrong-to-discourage-girls-from-being-competitive-1.3419556
5. phys.org/news/2010-07-competition-double-edged-sword-teenage-girls.html

PERSEVERANCE

1. coloradocac.com/ blog/10-ways-develop-perseverance-children/
2. today.com/parenting-guides/ building-perseverance-teens-t177667
3. psychologyeverywhere.com/articles/raising-girls-with-grit/
4. princeton.edu/news/2014/04/03/ perseverance-and-support-keys-womens-success-stem
5. selfsufficientkids.com/childrens-books-perseverance/

CONFIDENCE

1. theatlantic.com/family/archive/2018/09/ puberty-girls-confidence/563804/
2. childmind.org/ article/13-ways-to-boost-your-daughters-self-esteem/
3. canadianwomen.org/ blog/5-reasons-why-confidence-matters/
4. cnn.com/2018/05/21/health/girls-confidence-code-parenting/index.html
5. medium.com/the-motherload/girls-lose-confidence-when-theyre-only-eight-5561631d2928

SOFTWARE, GAMES, AND ORGANIZATIONS

Brown Girls Code—browngirlscode.org
Global nonprofit organization with programs to equip underrepresented girls with skills and training to pursue opportunities in Computer Science and other STEAM-related fields.

Code—Code.org
Non-profit organization dedicated to expanding participation in Computer Science by making it available in more schools and increasing participation by young women and students from other underrepresented groups.

Code Ninja—codeninjas.com
An educational organization specializing in teaching coding to kids, *way to introduce your kids to Computer Science and the world of programming.*

Girls Who Code—girlswhocode.com
International nonprofit organization that aims to support and increase the number of women in Computer Science in party by showing girls that coding is a problem-solving tool.

Hackergal—hackergal.org
Charity focused on tech education for girls and non-binary learners across Canada through online education, webinars, and hackathons.

Skills USA—skillsusa.org
Student-led partnership of education and industry. Using a framework, students hone skills including personal skills, workplace skills and technical skills grounded in academics and integrated into classroom curriculum.

Women Inspiring Girls and Women in STEM Excellence (WISE)—wise-stem.org
Non-profit Science, Technology, Engineering and Math (STEM) organization focused on helping Black and Brown women and girls in STEM using a holistic approach that includes tutoring, mentoring, technology and other support services.

ABCya—a website that provides educational games and activities for children in pre-kindergarten to sixth grade.

Animator—a platform for beginners, non-designers & professionals to create Animation and Live-Action videos for every moment of our life.

Boom Learning—a platform and set of tools for creating and assigning Boom Cards, cloud-based digital learning resources.

Discord—an instant messaging and VoIP social platform which allows communication thorough voice calls, video calls, text messaging, and media and files.

First Robotics Competition—a global program that challenges teams of high school students to build industrial-size robots and play a field game in alliance with other teams.

JavaScript—often abbreviated as JS, a programming language that is one of the core technologies of the World Wide Web.

Kleki—a free, open-source tool providing the ability to paint online with natural brushes, layers, and edit your drawings.

Kodable—learn about the fundamental of Computer Science with lessons ranging from zero to JavaScript using this game-based approach. Children can use creativity and critical-thinking skills to solve mazes as they learn.

Lego Blocks language—blocks of code including loops, logic, variables, Math, text and arrays snap into each other to define the program that will run.

Minecraft Education—engages students in game-based learning across the curriculum using a game-based platform that inspires creative, inclusive learning through play. Modules include coding, Math skills, Artificial Intelligence (AI) and Cyber Security. education.minecraft.net

Photoshop—a photo editing and raster graphic design software which allows users to create, edit and manipulate various graphics.

Pro-Create—an App that provides an Art Studio you can take anywhere to create expressive sketches, rich paintings, gorgeous illustrations and beautiful animations.

Python—a programming language that lets you work more quickly and integrate your systems more effectively.

Raspberry PI—a credit-card-sized minicomputer that turns a monitor, TV, mouse, or keyboard into a full-fledged PC.

Roblox—an online game platform and game creation system developed by Roblox Corporation that allows users to program games and play games created by other users.

Scratch—a free programming language and online community where you can create your own interactive stories, games and animations.

ScratchJr—program your own interactive stories and games for ages 5-7.

SPARK—digital studio app to create, inspire and play with others.

Subway Surfers—arcade game where you help Jake, Tricky and Fresh escape from the grumpy Inspector and his dog.

Toca Life—a series of games that encourage the player to imagine stories for characters in the game.

Virtual Reality (VR)—a simulated experience that employs pose tracking and 3D near-eye displays to give the user an immersive feel of a virtual world. Applications of virtual reality include entertainment, education and business.

WhatsApp—(officially WhatsApp Messenger) a freeware, cross-platform, centralized service to send text, voice messages and video messages, make voice and video calls, and share images and documents.

Acknowledgements

We thank our editors—Ann David, Lorraine Irwin, Jessica Keeling and Judy Prang. They believed in us, supported us and positively impacted our writing with their gentle hands. We appreciate the time and effort, and especially the talent they brought to polishing our book and bringing out our best.

We thank the parents who worked with us along the way, providing access to their daughters' stories and reviewing material to provide input on the messages they felt would resonate with girls exploring technology. Your time and perspective were a tremendous gift to us.

We thank the girls who shared their interest in tech, their stories, their engaging personalities and their talents with us. It is for you, and all girls, we author this book.

Kelley is thankful for the journey of writing about the terrific opportunities available for girls to explore a career in technology.

I remember being young and curious and discovering technology. I voiced my pleasure and couldn't believe people got paid to do something that was so much fun.

Talking to the girls and hearing their stories is uplifting. The women who have led with grace and strength inventing technology that impacts the world are inspirational. The women who are leading organizations to increase technology skills for females are trailblazers. Every woman succeeding in a technology role today is a visible reminder of the community ready to embrace our girls.

During the writing of this book, my dad was diagnosed with dementia. This impacted my life and the lives of my family and friends more than I could have expected.

I am blessed to have an extended family, circle of friends, and professional network who knew when I needed understanding, support, or simply encouragement to keep writing.

I am thankful my parents were the first people to hold a copy of the final draft of this book. Their smiles and joy are a gift to me.

I am grateful my husband Les never tired of my talking about this book or my plans to positively influence young girls.

Debra says this book is about empowering young girls, whether they choose a career in high tech or any other field. It is so important for them to be comfortable in their own skin, to believe in themselves, to be resilient, to try new things, and to be brave, bold, and sassy. I am describing my granddaughters who range in age from 3 to 17 years old.

I am blessed to have them in my world, to watch them grow, and step into whoever they are to be. I am impressed with their confidence in their own abilities, and their sense of self is well grounded. Their voices are strong, and they don't hesitate to let you know what they think or how they feel. I am encouraged when I see these girls and women stepping into the forefront in their own special way. They carry the strength and teachings of their mothers and grandmothers, all incredible women who carved out their own special place in the world.

I am enjoying every moment I get to spend in their company. They teach me something with every visit and interaction. They make me smile. They make me think. They make me pause as I watch them evolve. They bring me joy as I watch them grow and develop their unique talents and very special gifts.

They are the future; we need them, and I am proud to be their Grandy.

You make my heart swell. It beats a bit faster whenever I am in your presence. Thank you for being your wonderful selves.

About the Authors

Kelley Irwin is a Strategic Advisor and a Corporate Board Director, leveraging her experience as a technology executive. She leads with an intense sense of responsibility to create positive change.

She started her career as a software developer and has 35 years technology and executive management experience. Kelley was a Chief Information Officer (CIO) in both public and private companies.

She currently leads a practice providing Executive Bootcamps for technology start-up CEOs. She serves as a Board Director providing expertise in corporate governance, technology, and cyber security for companies in the technology, insurance, and utilities sectors.

Kelley is an active supporter of women in tech through mentoring, coaching, blogging, and speaking events.

She and her husband call the Toronto area home. They both enjoyed careers in technology and their daughter Emma and son Clayton have embraced technology as a career.

linkedin.com/in/kelleyirwin

www.kelleyirwin.com

Debra Christmas is a 45-year veteran of the information technology industry. She spent almost three decades with a leading technology vendor in consulting, sales, channels, and software. She was a Chief Information Officer in the municipal sector, driving change and transformation in the delivery of constituent services to citizens in Ontario.

She is currently a Senior Executive Partner at Gartner Canada Inc. and has the honour and privilege of working with Senior Technology Executives as they lead their organization through change, driving innovation and leveraging the power of technology assets. She has travelled the world as a facilitator for Manager Milestones, an innovative leadership development program and recipient of the Brandon Hall Gold Excellence Award for Learning. She has engaged with hundreds of leaders across continents and cultures to help them develop leadership skills to drive business results and enhance employee engagement. She is the Co-Creator of PACE an award-winning presentation and communications program.

Debra is a trained Co-Active and John Maxwell Leadership Coach and is certified in Tilt365 and Leadership Circle, a comprehensive leadership assessment framework based on "Mastering Leadership". She is the founder of Stiletto Gladiators and has been active in equity, diversity, and inclusion initiatives for women her entire career. She will never give up the fight.

linkedin.com/in/debrachristmas

www.stilettogladiators.com

Please Join Us

We are opening the door for a dialogue, an exchange of ideas, a conversation about this profession, the jobs and careers available, and the impressive tribe of people you could join in the journey.

Visit us at: www.womenintechtribe.com to hear our ongoing stories and perspectives.

Join us at: https://www.linkedin.com/company/womenintechtribe to contribute with your stories, your words of inspiration, your questions and your thoughts. We want to hear your interests, your concerns and of course your successes. We can teach, and we can learn. And isn't that what a great community is all about?

Women in tech are powerful. Join the discussion; join the movement.

It is time for us to unite, to come together as a community of smart, capable, competent girls and women in tech. There is strength in numbers. There is power with focus and commitment to a cause.

Girls in tech are the cause. We have the right to succeed and the responsibility to help others.

Endnotes

1. Women Tech Network, 2023, https://www.womentech.net/en-ca/women-technology-statistics
2. Forbes, 2017, www.forbes.com/sites/lisawang/2017/12/08/the-surprising-reason-girls-are-not-getting-into-tech/?sh=3956bdd6f780
3. Science Daily, 2021, https://www.sciencedaily.com/releases/2021/11/211122172716.htm
4. Anouk Wipprecht FashionTech, 2023, www.anoukwipprecht.nl
5. Karlie Kloss | Supermodel & Entrepreneur, 2023, www.karliekloss.com
6. Erika Rawes, Digital Trends, 2022, www.digitaltrends.com/cool-tech/influential-women-in-tech
7. Erika Rawes, Digital Trends, 2022, www.digitaltrends.com/cool-tech/influential-women-in-tech
8. Wikipedia, 2023, https://en.wikipedia.org/wiki/Reshma_Saujani
9. Ypulse, 2022, https://www.ypulse.com/article/2022/11/14/young-girls-are-losing-confidence-but-mentorship-could-change-that
10. biography.com, 2021, www.biography.com/scientist/grace-hopper
11. encyclopedia.com, 2019, https://www.encyclopedia.com/history/encyclopedias-almanacs-transcripts-and-maps/mary-g-ross
12. Britannica, 2023, https://www.britannica.com/biography/Mary-Golda-Ross

13. Science Daily, 2021, sciencedaily.com/releases/2015/08/150817085619.htm
14. businesschief, 2022, https://businesschief.asia/leadership-and-strategy/top-10-women-in-technology-in-asia-pacific
15. Goldhouse, 2024, https://goldhouse.org/people/neha-parikh/
16. Carl Linberg, Leadership Ahoy, 2022, https://www.leadershipahoy.com/the-six-leadership-styles-by-daniel-goleman/
17. Queen Elizabeth Prize for Engineering Foundation, 2023, https://qeprize.org/nominate/judges
18. Public, 2023, https://www.publicnow.com/view/C97EE066FDF0AF41913F7082F4143A7DF9504F1E
19. ACCN, 2024, https://accntoronto.com/)
20. Soulpepper, 2024, https://www.soulpepper.ca/
21. Wikipedia, 2023, https://en.wikipedia.org/wiki/Ursula_Burns
22. Computer History Museum, 2008, https://computerhistory.org/profile/jean-jennings-bartik/
23. Wikipedia, 2024, https://en.wikipedia.org/wiki/Jean_Bartik
24. FIRST LEGO League, 2023, www.firstlegoleague.org
25. Wikipedia, 2024, en.wikipedia.org/wiki/Annie_Easley
26. Wikipedia, 2023, en.wikipedia.org/wiki/Mary_Allen_Wilkes
27. Mindkiss, the Project, 2023, www.mindkiss.com

28. Erika Rawes, Digital Trends, 2022, www.digitaltrends.com/cool-tech/influential-women-in-tech
29. Erika Rawes, Digital Trends, 2022, www.digitaltrends.com/cool-tech/influential-women-in-tech
30. Wikipedia, 2023, https://en.wikipedia.org/wiki/Karen_Sp%C3%A4rck_Jones
31. Igor Ovsyannnykov, Inspirationfeed, 2023, https://inspirationfeed.com/inventor-of-voip/
32. USBE Information Technology, 2018, https://www.blackengineer.com/article/dr-gladys-west-inducted-in-pioneers-hall-of-fame/